Internet of Things

Technology, Communications and Computing

The series Internet of Things - Technologies, Communications and Computing publishes new developments and advances in the various areas of the different facets of the Internet of Things. The intent is to cover technology (smart devices, wireless sensors, systems), communications (networks and protocols) and computing (theory, middleware and applications) of the Internet of Things, as embedded in the fields of engineering, computer science, life sciences, as well as the methodologies behind them. The series contains monographs, lecture notes and edited volumes in the Internet of Things research and development area, spanning the areas of wireless sensor networks, autonomic networking, network protocol, agent-based computing, artificial intelligence, self organizing systems, multi-sensor data fusion, smart objects, and hybrid intelligent systems.

Indexing: *Internet of Things* is covered by Scopus and Ei-Compendex **

Kashif Naseer Qureshi
Thomas Newe • Gwanggil Jeon
Abdellah Chehri
Editors

Cybersecurity Vigilance and Security Engineering of Internet of Everything

Springer

Editors
Kashif Naseer Qureshi
Department of Electronic and Computer Engineering
University of Limerick (UL)
Limerick, Ireland

Thomas Newe
Department of Electronic and Computer Engineering
University of Limerick (UL)
Limerick, Ireland

Gwanggil Jeon
Department of Embedded Systems Engineering
Incheon National University
Incheon, Korea (Republic of)

Abdellah Chehri
Department of Mathematics and Computer Science
Royal Military College of Canada
Kingston, ON, Canada

ISSN 2199-1073 ISSN 2199-1081 (electronic)
Internet of Things
ISBN 978-3-031-45161-4 ISBN 978-3-031-45162-1 (eBook)
https://doi.org/10.1007/978-3-031-45162-1

This Springer imprint is published by the registered company Springer Nature Switzerland AG
The registered company address is: Gewerbestrasse 11, 6330 Cham, Switzerland

Paper in this product is recyclable.

Editor's Note

In today's hyperconnected world, where technology permeates every aspect of our lives, the Internet of Everything (IoE) has emerged as a transformative force. With billions of devices connected to the internet, ranging from household appliances to industrial machinery, people, processes, data, and things to modernize their activities. The IoE has brought unprecedented convenience and efficiency. However, it has also exposed us to new and complex cybersecurity challenges. *Cybersecurity Vigilance and Security Engineering of Internet of Everything* is a timely and essential book that delves into the intricate world of securing the IoE. As we continue to witness the rapid proliferation of connected devices, it is crucial to understand the threats and vulnerabilities that arise in this interconnected ecosystem. This book serves as a comprehensive guide for individuals and organizations seeking to navigate the ever-evolving landscape of IoE security.

The authors of this book, with their deep expertise and extensive experience in the field of cybersecurity, provide valuable insights into the unique risks associated with the IoE. They meticulously analyze the diverse range of devices, networks, and applications that constitute the IoE, shedding light on the potential entry points for malicious actors and the devastating consequences of successful attacks. By presenting real-world case studies and practical examples, they offer actionable strategies for detecting, preventing, and mitigating cyber-threats in an IoE environment. Cyber-attacks on information systems in IoE networks are a crucial area of research and need detailed investigation.

This book suggests two parts of security fundamentals including security threats and vulnerabilities, security vigilance, and security engineering for IoE networks and goes beyond mere theoretical concepts, equipping readers with the necessary knowledge and tools to proactively address cybersecurity concerns. The book explores cutting-edge technologies, such as artificial intelligence and blockchain, and their application in enhancing IoE security. It emphasizes the importance of adopting a holistic approach to cybersecurity, encompassing not only technical measures but also organizational policies, user awareness, and regulatory frameworks. As the digital landscape continues to evolve, cybersecurity vigilance becomes paramount. The interconnected nature of the IoE presents both immense opportunities and profound risks.

This book serves as a beacon of knowledge and guidance, empowering readers to safeguard their critical systems, protect their privacy, and contribute to the overall security of the IoE ecosystem. I commend the authors for their comprehensive research, diligent analysis, and commitment to advancing cybersecurity in the context of the IoE. Their work will undoubtedly make a significant contribution to the field and will serve as a valuable resource for cybersecurity professionals, researchers, and policymakers alike. I encourage readers to delve into the pages of *Cybersecurity Vigilance and Security Engineering of Internet of Everything* and embark on a journey toward understanding the intricate challenges and developing robust solutions to secure our increasingly interconnected world.

I believe that *Cybersecurity Vigilance and Security Engineering of Internet of Everything* will be useful to readers who are starting to approach this complex technical topic, since it puts together many different perspectives, application examples, and specific solutions. At the same time, it will be a useful reference for the more experienced researcher who aims at going deeper into a specific vertical application of IoE networks, or who looks for possible open questions and/or future research topics to be explored.

Department of Electronic & Computer Engineering
University of Limerick (UL),
Limerick, Ireland

Kashif Naseer Qureshi

Preface

Internet of Everything (IoE) technology is based on intelligent connections among people, processes, data, and things to modernize their activities. Cyberse-curity is one of the main challenges for these networks due to the specific charac-teristics of the network and the weak nature of connected things. The billions of objects are connected over public and private networks and expose these net-works to security risks and breaches. Cyber-attacks on information systems in IoE networks are a crucial area of research and need detailed investigation. In the context of IoE networks, cybersecurity encompasses the formulation of policies, utilization of tools, application of security concepts, implementation of detection and prevention mechanisms, deployment of security safeguards, protection of user assets, assurance measures, and incorporation of relevant technologies. This book suggests two parts of security fundamentals including security threats and vulnerabilities, security vigilance, and security engineering for IoE networks. The first section covers the security threats and vulnerabilities or techniques to expose the networks to security attacks such as repudiation, tampering, spoofing, and el-evation of privilege. The second section of the book covers vigilance or preven-tion techniques like intrusion detection systems, trust evaluation models, crypto, and hashing privacy solutions for IoE networks. This section also covers the secu-rity engineering for embedded and cyber-physical systems in IoE networks such as blockchain, artificial intelligence, and machine learning-based solutions to se-cure the networks. This book provides a clear overview in all areas to understand the readers about IoE networks in terms of security threats, prevention, and other security mechanisms.

This book entails two sections and twelve chapters, including the following studies.

Part A: Security Threats and Vulnerabilities

In Chap. 1, Kashif Naseer Qureshi et al. discussed the Internet of Everything: Evolution and fundamental concepts, technologies and applications. In addition, the authors proposed a layer architecture for IoE networks adoption and sustainability.

In Chap. 2, Bahareh Pahlevanzadeh, and Sara koleini highlighted the new cyber-security risks, threats and attacks in IoE networks. In addition, authors also proposed a 3D cy-bersecurity model based on the presented IoE components, enabler technologies, and multi-layered architecture.

In Chap. 3, Raja Waseem Anwar and Kashif Naseer Qureshi highlighted the fundamental concepts of existing attack detection mechanisms adopted at the edge and cloud-based networks to secure the IoE networks. This chapter also explores the back-bone network threats especially edge and cloud computing or other service pro-viders for IoE networks.

In Chap. 4, Hilary Meagher and Lubna Luxmi Dhirani discussed the various potential threat scenarios (i.e., espionage, loss of command and control, compromised data or de-vice, etc.), high-risk concerns that need to be considered for building cyber-resiliency for IoE networks. Best practices, standards, risks, policies and alignment with cyber-resilience act and law are also summarized in this chapter.

In Chap. 5, Saleem Iqbal et al. discussed the future cybersecurity challenges and demand for new and integrated security architecture for IoE network and explored the securi-ty architecture requirements from application's perspective.

Part B: Security Vigilance and Security Engineering for IoE Networks

In Chap. 6, Fasee Ullah and Asad Ullah presented the network and security architecture for IoE networks and explored the security architecture requirements for each appli-cation.

In Chap. 7, Amna Khatoon et al. discussed the machine and deep learning-based detection and prevention methods for IoE networks along with their advantages and disad-vantages to protect the networks from unknown attacks.

In Chap. 8, Ibrahim Tariq Javed and Kashif Naseer Qureshi discussed the role of blockchain models for IoE infrastructures and applications. In addition, a framework is also suggested for creating blockchain-based IoE application solutions.

In Chap. 9, John Morris et al. discussed the cybersecurity as a service including common CSaaS functions and their providers. Moreover, chapter also explored the guidance especially for small- and medium-sized businesses, for asking the appropriate questions when it comes to the selection of a specific MSSP.

In Chap. 10, Faisal Rehman et al. discussed the big data analytics for cybersecurity in IoE networks. The comprehensive details are also added to understand that how big data analytics can be applied to the task of creating a trustworthy IoE.

In Chap. 11, Kashif Naseer Qureshi et al. presented the cybersecurity standards and policies for CPS in IoE. This chapter also discusses the security vulnerabilities and privacy threats of Cyber-Physical Systems (CPS) in IoE networks and presented security and privacy solutions/architectures that improve the security and privacy of CPS in IoE networks.

In Chap. 12, Abeer Iftikhar and Kashif Naseer Qureshi discussed the privacy, and trust chal-lenges for IoE networks This chapter also highlighted the new trend and usage of artificial intelligence in IoE networks.

This book is designed for researchers, engineers, and developers working in the fields of IoE networks with emerging technologies like 5G/6G, security standards, and blockchain. Practitioners who conduct teaching and cutting-edge research in secure IoE environments will be benefited from this book. Special thanks to all contributors, respected referees, and our publisher, Springer.

Limerick, Ireland — Kashif Naseer Qureshi
Limerick, Ireland — Thomas Newe
Incheon, Korea (Republic of) — Gwanggil Jeon
Kingston, ON, Canada — Abdellah Chehri

Acknowledgments

This work received support from the Higher Education Authority (HEA) under the Human Capital Initiative-Pillar 3 project, Cyber Skills.

Cyber Skills

Cyber Skills is a HEA Human Capital Initiative-funded under pillar 3 that brings together Ireland's leading experts in cybersecurity education. Munster Technological University, University of Limerick, and Technological University Dublin are collaborating to provide pathways and micro-credentials to address skill shortages in the area of Cyber Security. The programs are designed to provide industry-based learners with the knowledge and skills designed to enhance their careers with cybersecurity expertise.

Cyber Skills is the only resource in Ireland where you will find courses and modules that have been specifically designed and created by industry and academic experts. Working closely with our industry partners, including Dell, Mastercard, ADI, J&J, etc., Cyber Skills have designed courses informed by the needs of the workplace to enhance the skills of IT Cybersecurity professionals in a wide variety of industries.

Cyber Skills benefits from the first-of-its-kind world-class cloud-based Cyber Range and mobile Cyber Range unit. These Cyber Ranges provide a secure, sandboxed area which simulates real-world feel scenarios and environments where students can test their new skills in an environment that can replicate their work-based systems. Labs and assignments will be used to reinforce the content from the lectures and a full range of scenarios will provide the opportunity to test the vast array of techniques required to keep ahead in this challenging and ever-changing environment.

Since its foundation, Cyber Skills has recruited over 14 academic staff who come from multi-disciplinary backgrounds, with a passion for cybersecurity. Combining years of experience with expert knowledge, our lecturers enable students to achieve their academic and career progression goals.

All programs delivered by Cyber Skills are aligned to the Internationally recognized NIST-NICE Cybersecurity Workforce Framework. This framework allows Cyber Skills to provide clear guidance to learners on the work role tasks, knowledge, and skills being covered in their selected course so that they may select a course that suits the work role they are in or planning to apply for.

Contents

About the Authors

Kashif Naseer Qureshi is an Associate Professor of Cyber Security in the Department of Electronic and Computer Engineering at the University of Limerick, Ireland. He is also actively involved in the Cyber Skills project, a HEA-HCI Pillar 3 initiative Ireland. He received a Ph.D. degree from the University of Technology Malaysia (UTM) and holds two master's degrees in Computer Science and Information Technology from reputable universities. He is the Co-Principal Investigator in Cyber Reconnaissance and Combat project funded by higher education commission. His research interests focus on the security, trust and privacy concerns for Internet of Everything (IoE), Internet of Vehicles (IoV), Electronic Vehicles (EV) charging management planning and recommendation systems, and Internet of Things (IoT) and use cases implementation in wireless and wired networks. He is active member of Lero, the Science Foundation Ireland Research Centre for Software in University of Limerick (UL). His name is included in top 2% Scientist for consecutive 3 years from Stanford university USA. He has published various high-impact factor papers in international journals and conference proceedings and served on several conferences IPCs and journal editorial boards. He has number of book chapters and 5 edited books in Springer, CRC and Elsiver Publishers related to Cybersecurity, Privacy and Trust architectures. He has also part of various research projects related to wireless communication, routing and CyberSecurity domains in the UK, China, Ireland, Malaysia, Canada, Dubai, Vietnam and in Pakistan.

Thomas Newe is an Associate Professor of Cyber Security in the Department of Electronic and Computer Engineering at the University of Limerick and is the Principal Investigator in the SFI Smart Manufacturing Centre, Confirm and a Funded Investigator in the SFI Centres, Lero-Software Research Centre, and MaREI-Marine and Renewable Energy Research Centre. He holds a B.Eng. in Computer Engineering, a Master in Engineering in Security Protocol Design, and a Ph.D. in Formal Logic for Security Protocol Verification. He has been a University of Limerick faculty member since 1994. Tom is a board member of Cyber Ireland, an initiative that brings together Industry, Academia, and Government to represent the needs of the Cyber Security Ecosystem in Ireland, and a founding member of Cyber Skills, a HEA-HCI Pillar 3 initiative that aims to address the global shortage of cybersecurity professionals. His research interests include many topics under the general areas of cyber security for Data, Networks, the Internet of Things, and Smart Collaborative Robotics. He has to date graduated 16 Ph.D. students in these areas.

Gwanggil Jeon received the B.S., M.S., and Ph.D. (summa cum laude) degrees from the Department of Electronics and Computer Engineering, Hanyang University, Seoul, Korea, in 2003, 2005, and 2008, respectively. From September 2009 to August 2011, he was with the School of Information Technology and Engineering, University of Ottawa, Ottawa, ON, Canada, as a Post-Doctoral Fellow. From September 2011 to February 2012, he was with the Graduate School of Science and Technology, Niigata University, Niigata, Japan, as an Assistant Professor. From December 2014 to February 2015 and June 2015 to July 2015, he was a Visiting Scholar at Centre de Mathématiques et Leurs Applications (CMLA), École Normale Supérieure Paris-Saclay (ENS-Cachan), France. From 2019 to 2020, he was a Prestigious Visiting Professor at Dipartimento di Informatica, Università degli Studi di Milano Statale, Italy. He is currently a Full Professor at Incheon National University, Incheon, Korea. He was a Visiting Professor at Sichuan University, China, Universitat Pompeu Fabra, Barcelona, Spain, Xinjiang University, China, King Mongkut's Institute of Technology Ladkrabang, Bangkok, Thailand, and the University of Burgundy, Dijon, France. Dr. Jeon is an IEEE Senior Member, and Associate Editor of *Sustainable Cities and Society*, *IEEE Access*, *Real-Time Image Processing*, *Journal of System Architecture*, and *MDPI Remote Sensing*. Dr. Jeon was a recipient of the IEEE Chester Sall Award in 2007, the ETRI Journal Paper Award in 2008, and the Industry-Academic Merit Award by the Ministry of SMEs and Startups of Korea Minister in 2020.

Abdellah Chehri is passionate for technology, innovation, learning, and teaching. He received his M.A.Sc. degrees in Signal and Digital Communication from the University Sophia-Antipolis at Nice-France. In September 2004, he joined the Department of Electrical and Computer Engineering of Laval University, Quebec, Canada, where he was a member of the Radio communications and Signal Processing Laboratory (LRTS). He received his Ph.D. in Electrical Engineering in June 2009 under the supervision of Prof. Paul Fortier and Dr. Pierre Martin Tardif. From 2007 to 2009, he worked as a project member at the Bell-Aliant Research Laboratory, Quebec, Canada. During this period, he also served as a Lecturer in the Information Technologies Graduate Program of UQAT. Dr. Chehri received a number of prestigious awards, including Dean's Scholarship Award, Postdoctoral Studies (University Ottawa), Scholarship Fund to Support Success (Laval University), Japan Society for the Promotion of Science, MITACS, NSERC Postdoctoral Fellowship. He joined the University of Ottawa in July 2009 as a Post-doc Fellow/Research Associate and Member of Wireless Heterogeneous Sensor Networks in the e-Society WiSense project working under the supervision of Prof. Hussein. T. Mouftah. From 2012 to

2014, he was with BLiNQ Networks in Kanata, where he was involved in the development and testing of dual-carrier, joint scheduling, interference avoidance, and beam selection algorithms for NLOS TDD backhaul for small-cell 4G/LTE mobile applications. He is working as an Adjunct Assistant Professor at the School of Electrical Engineering and Computer Science of the University of Ottawa (Electronics, Signal Processing, Telecommunications, and Electromagnetism). Dr Chehri has published more than 70 research papers and a number of book chapters. Dr. Chehri served as the program chair and workshops chair of the 2014 IEEE International Humanitarian Technology Conference (IEEE IHTC 2014). He served as guest editor and technical reviewer for several international conferences and journals (IEEE, Elsevier, Springer). Additionally, he is a Senior Member of IEEE Communications Society and IEEE-Ottawa Section, and Member of the Order of Engineers of Quebec (OIQ) and IEEE Canadian Humanitarian Initiatives Committee (HIC).

Part I
Security Threats and Vulnerabilities

Chapter 1
Internet of Everything: Evolution and Fundamental Concepts

Kashif Naseer Qureshi, Thomas Newe, Gwanggil Jeon, and Abdellah Chehri

1.1 Introduction

Internet of Everything (IoE) is one of the new concepts and new versions of the Internet of Things (IoT) where everything is connected to the Internet and offers several sensing and monitoring services [1]. The difference between these two technologies is their connections and ability to analyze the data. The IoT network refers to a connection of physical devices such as sensor nodes, communication devices, and other Internet wearable smart gadgets. These devices are communicating with each other for data sharing without human intervention. On the other hand, the IoE network refers to a connection between not just physical devices but also people, data, and processes. In these networks, the devices and other elements are communicating to capture and analyze data in real time for better decisions. The main aim of IoE is to create a fully interconnected intelligent system to sense, analyze, and respond to users' requirements. The implementation of IoE networks depends on communication systems, low-power communication standards, and big data analytics systems. The new advancements and breakthroughs in technologies open new

K. N. Qureshi (✉) · T. Newe
Department of Electronic & Computer Engineering, University of Limerick (UL), Limerick, Ireland
e-mail: Kashifnaseer.qureshi@ul.ie; thomas.newe@ul.ie

G. Jeon
Department of Embedded Systems Engineering, Incheon National University, Incheon, Korea
e-mail: gjeon@inu.ac.kr

A. Chehri
Department of Mathematics and Computer Science, Royal Military College of Canada, Kingston, ON, Canada
e-mail: chehri@rmc.ca

K. Naseer Qureshi et al. (eds.), *Cybersecurity Vigilance and Security Engineering of Internet of Everything*, Internet of Things,
https://doi.org/10.1007/978-3-031-45162-1_1

doors for smart and intelligent systems like Artificial Intelligence (AI) [2]. The concept of IoE was introduced in 2012 by CISCO to envision the promising future of the Internet. In the first stage, IoE introduced four main pillars including people, data, processes, and things, whereas the IoT only contains things. The IoE network extended the concept of people-based processes for all things and enriched people's lives by enabling automated systems. There is a wide range of IoE applications starting from manufacturing, intelligent transportation systems, smart homes, smart agriculture, and smart healthcare services.

IoE networks aim to provide real-time communication services for everyday objects like home appliances, vehicles, and industrial machines. These networks would allow effective way and resources utilization and new advance controlling processes. Sensor nodes, embedded systems, cloud and edge computing services, data analytics, and advanced technologies are used for data communication. One of the significant advantages of IoE networks is analyzing and gathering a large amount of data and optimizing and improving the processes. In healthcare, smart devices are used to monitor patient data and alert the medical staff before any emergency. In the smart industrial section, the IoE devices are used to decrease downtime, fault monitoring, and improve productivity [3]. In intelligent transportation systems, the IoE networks improve safety, especially in smart traffic management systems to control traffic congestion issues. The most popular application for transportation systems is accident alert systems, road condition management, and navigation services.

With many features, the IoE networks have suffered from several intrinsic limitations like network coverage, network access, battery constraints, and security issues. The network coverage especially in harsh and remote geographical areas is one of the limitations of these networks due to the unavailability of infrastructure and its deployment. The energy of small nodes is another considerable constraint of IoE networks due to portability considerations. The sensor nodes are easily exhausted and lead to connection loss, delay, and network overhead. The vision of the IoE network is consideration of new protocols like Internet Protocol version 6 (IPv6) and extend this network with more smart Augmented Reality (AR) and AI capabilities. Virtualization data services are also considered for IoE networks. The vision of IoE networks is to analyze the massive data generated from connected devices. The three expectations of IoE devices are scalability, intelligence, and diversity. Scalability refers to a scalable network that covers all data communication requirements in all geographical areas such as in urban, rural, and terrestrial space. The IoE networks need wide coverage and communication standards for long- and short-range communication. These networks are also integrated with cellular networks, satellite, and mobile ad hoc networks. The scalability also supports the physical data collection for data analysis. Intelligence is another requirement and vision of IoE networks where decisions, predictions, and other actions are applied to collected data. Diversity is another requirement of IoE networks due to different applications for people-based processes and automation.

1.2 Enabling Communication Technologies for IoE Networks

The communication technologies for IoE networks enable massive access to devices and coverage for networks. The new and advanced IoE networks need different types of support in terms of distance, and network topologies. To achieve the IoE network requirements, the technologies are divided into three main areas including backbone technologies, local handling technologies, and end-user technologies. The backbone technologies offer an overall connection for data transmission, collection, and processing. The backbone technologies are Mobile Communication Networks (MCN) and Low-Power Wide Area Networks (LP-WANs) [4]. The MCN technologies cover dense ad crowd areas like urban areas. The IoE networks access the backbone networks by using the 5G and 6G technologies. On the other hand, the LP-WANs are used for massive IoE nodes, especially for smart agriculture and forest monitoring. The local technologies refer to those which are working from short distances like a few meters such as Zigbee, 6LoWPAN, and Bluetooth technologies. Wi-Fi is one of the wireless technology for wireless communication. Cellular networks are utilized for wireless connections, especially for hard-to-reach areas, and used for mobile network operators. Bluetooth is used for short-range data communication and is commonly used for IoE devices like smartwatches, wearable devices, and smartphones. Radio Frequency Identification (RFID) is used in IoE networks for identifying and tracking objects for asset tracking and supply chain management [5]. Near Field Communication (NFC) is used for short-range communication such as for access control and contactless payments. Figure 1.1 shows the three-layer-based communication technologies for IoE networks.

Various new emerging technologies like Artificial Intelligence (AI), blockchain technologies, and quantum computing offer new application scenarios and the final development of IoE networks. AI technologies are based on new algorithms and human thinking technologies to solve complex problems in IoE networks. The achievements of AI are well recognized in different areas and fields such as natural language processing, computer vision, and intelligent robots. Deep Learning (DL) technologies are used in AI like pattern recognition and analyzing and detecting abnormal data. Blockchain technologies are also used in IoE networks, especially for privacy and security concerns. This technology provides a decentralized mechanism without relying on one central system. Blockchain technology also provides transparency in transactions and performs large-scale computing collaboration among IoE networks, edge/cloud, and backbone networks.

1.3 IoE Applications

The IoE network has the potential to transform and change traditional networks by introducing interconnected networks. With the development of IoE, it is quite unpredictable to consider its applicability in different dimensions. IoE is an exciting

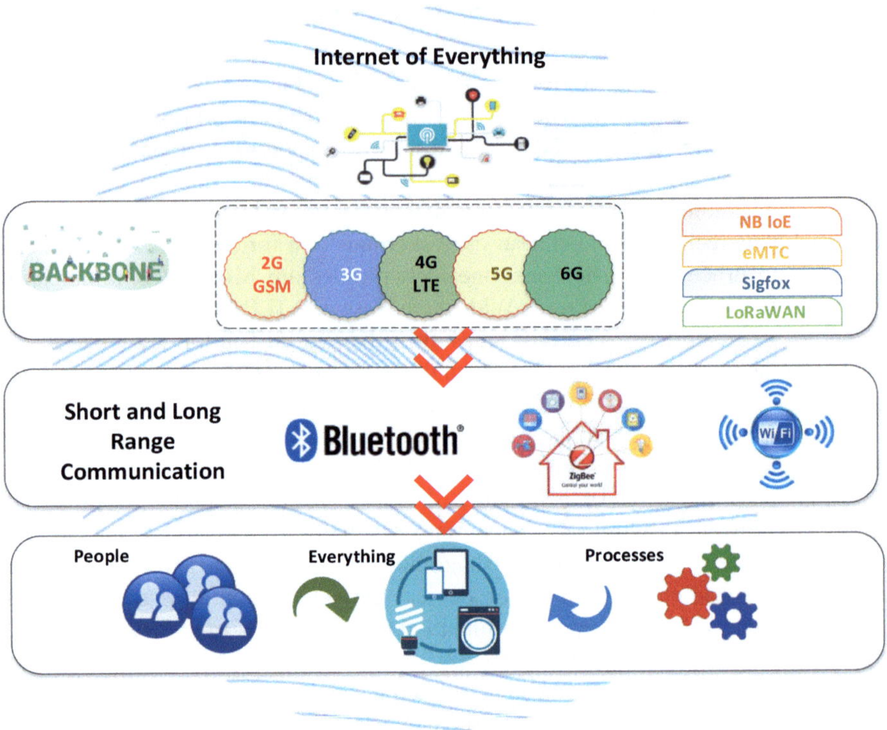

Fig. 1.1 Three-layer-based communication technologies for IoE networks

and rapidly evolving field that will shape and change the future of various industries in more efficient and novel ways in the coming years. Based on current research in various search engines and websites, various potential use cases of IoE applications are categorized by different sectors as follows:

- *Smart Grids:* Smart grid systems provide modern electrical grid systems to manage electricity transmission, generation, and distribution. These networks also provide advanced metering infrastructure for real-time energy usage to monitor electric consumption. IoE networks are used for smart grids to control and monitor the electricity management systems for optimized energy production and consumption. IoE-connected smart meters are working with real-time energy usage monitoring and optimization. Predictive maintenance of energy infrastructure using real-time data analytics and automated energy management systems for homes and buildings is a more attractive IoE application for users. Renewable energy management and optimization using IoE-connected sensors and data analytics are also gaining users' attention.
- *Smart Homes:* These networks provide smart appliance control systems to manage the devices remotely by using Internet services. These networks provide cost-effective management services for home appliances such as light control,

camera control systems, and other sensor nodes to control smoke, fire, and other home-related issues. All the smart home appliances are controlled via smartphones to adjust the thermostat and start and off the machines.

- *Smart Healthcare Systems:* These systems provide remote patient monitoring systems to control and record all vital signs of patients before any emergency. Blood pressure, temperature, and heartbeat record are common applications of smart healthcare systems. The smart sensor nodes are implanted inside or outside the patient's body to monitor the patient. IoE can be used to monitor patients' health in real time and alert doctors and caregivers to any issues. Wearable devices such as fitness trackers and smartwatches can track vital signs and send alerts if there are any changes. The IoE devices monitor the patient and manage the sensed data through connected medical devices. Smart medication management systems are also attractive services for patients and healthcare providers. IoE-enabled telemedicine services for remote consultations and diagnoses and smart hospital equipment and facilities management to improve patient outcomes are a few more examples of IoE network applications and services. Wearable technology for health and fitness monitoring for personalized care is also beneficial for human health.
- *Intelligent Transportation Systems:* IoE can be used to create smarter transportation systems, from autonomous vehicles for smart traffic management. Connected cars can communicate with each other and with the infrastructure to avoid accidents and reduce congestion. Real-time tracking and monitoring of goods is another service for transit by using IoE-connected devices. Optimized route planning and vehicle maintenance for fleets using data analytics and smart traffic management systems using real-time data from sensors and cameras are well-known services. Predictive maintenance and repair of vehicles and equipment and usage of autonomous vehicles and drone technology are used for safe and efficient transportation.
- *Smart Agriculture Systems:* IoE can be used to optimize crop yields and reduce waste. Sensors can monitor soil moisture, temperature, and other variables to ensure that crops receive the right amount of water and nutrients. Precision farming by using sensors and drones for optimized resource usage is one of the beneficial applications. Real-time weather monitoring for optimized crop management improves the crop production process. Livestock monitoring and management using IoE-connected sensors are attractive applications for farmers. Soil health and nutrient management through IoE-connected sensors and analytics and smart irrigation systems with real-time monitoring and optimization are the most attractive IoE applications.
- *Smart Industrial Systems:* IoE can be used to improve efficiency and reduce downtime in factories. Connected machines can communicate with each other and with the cloud to detect and diagnose problems before they become serious. Predictive maintenance for machinery by using real-time data from sensors and asset tracking and monitoring for supply chain optimization are examples of IoE network operations in industries. Quality control and defect detection using machine learning and computer vision are another area to improve productivity

and business. Smart inventory management with real-time monitoring and alert system is utilized in industries. Optimized energy consumption and resource management in factories and warehouses are more economical services of IoE networks in industries.

- *Retail and Hospitality Sector:* Personalized shopping experiences with IoE-enabled beacons and sensors are the most common application. Real-time inventory management with IoE-connected devices and analytics improves the retail and hospitality business. Smart vending machines and kiosks with real-time data analytics are used for optimized operations. Smart hotel rooms and guest experiences using IoE-connected devices and services are the most attractive applications of IoE networks. Predictive maintenance for hospitality equipment and facilities by using real-time data is also gaining users' attention.
- *Smart Cities and Infrastructure Sector:* Smart street lighting for energy efficiency and optimized maintenance has been adopted to improve the citizen lifestyle in urban areas. Real-time air quality monitoring and management for pollution control is another beneficial application. Intelligent waste management systems and smart building management systems for optimized energy consumption and maintenance are used to improve life quality. Automated parking management systems for optimized space usage and reduced congestion in urban cities.
- *Financial Services Sector:* Fraud detection and prevention systems using real-time data analytics and machine learning are used in the financial services sector. Real-time risk management and investment decision-making using IoE-connected devices improve business processes. Personalized financial advice and services using IoE-enabled devices and platforms are attractive applications. Automated wealth management and portfolio optimization using real-time data analytics and IoE-enabled payment and banking services for secure and efficient transactions are gaining more attention due to their fast and intelligent service provision.
- *Smart Education Systems:* Smart classroom management systems with real-time monitoring and analytics are used for personalized learning experiences with IoE-enabled devices and platforms. Smart campus management systems for optimized resource usage and maintenance are also examples of smart education systems. Real-time tracking and monitoring of student attendance and performance to improve the education system quality are one of the beneficial services. IoE-enabled remote learning and collaboration platforms for global education are significant after a long pandemic where every single institute has shifted to online platforms.
- *Entertainment and Media Sector:* Personalized content delivery and recommendations with IoE-connected devices are used in media and entertainment platforms. Smart advertising and marketing campaigns with real-time data analytics improve media coverage and user engagement. IoE-enabled gaming and virtual reality experiences for immersive entertainment are more advanced services of IoE. Real-time audience engagement and feedback using IoE-connected devices are also used to improve the process's quality.

These are just a few examples of the many applications of IoE. As technology continues to develop, we can expect to see even more innovative uses of connected devices and data analytics.

1.4 IoE Execution and Implementation Challenges

The development and implementation of OpenIoE systems are associated with several deployments, scalability, infrastructure, and standards issues and challenges. The execution of IoE networks is still under consideration in different fields due to different technologies integration such as edge computing, AI, blockchain, security, and semantic web aspects. The transformation and adoption of every (devices, sensors) data management for a common semantic format are two of the challenges for researchers and companies. Several different components and elements are involved where resources are needed to deliver the requested services. The implementation of middleware for IoE networks is still a challenge due to physical virtual things or sensor involvement. Big data handling and its process management are other significant areas of research in IoE networks. The applications integration with cloud and edge networks also needs attention for a better integrated development environment.

The IoE project execution at small and large scales is another challenge due to the integration of components and heterogeneity of everything. The traditional IoT networks are using cloud computing where the data is collected from terminal devices for further processing. The large-scale data processing degraded traditional cloud-based services in terms of latency, bandwidth, and data processing. The IoE networks are utilizing edge computing closer to the IoE networks for better services. However, edge networks are under consideration due to diverse architectures and standards. Security and user privacy are other challenges due to the open nature of networks. There are higher chances for cyberattacks in IoE networks due to more connected devices with other networks for data communication. Interoperability is another challenge because different companies and manufacturers are using different protocols and standards. Data management and complexity are other challenges of IoE network adoption due to a large amount of data. There is a need for advanced data analytic modes and methods to handle the sensed data and extract useful information. Social and ethical implication raises in terms of user privacy, inequality, and autonomy which need attention to realize the full potential of IoE networks. The IoE networks are beneficial for organizations or industries if the following challenges related to execution and implementation will be resolved.

- *Scalability*: Due to several IoE devices and the involvement of different manufacturers and companies, the scalability and issue were raised for reliable connectivity in existing systems and networks.
- *Interoperability*: The different devices and systems lead to interoperability issues in the network due to different communication standards and protocols.

- *Security and Privacy*: IoE networks create new security threats and violations due to potential attack surfaces. Security and privacy provision is crucial due to heterogeneous and multi standards-based networks.
- *Data Management*: Data management and analytics are used for data analysis, collection, and management. This is one of the significant challenges due to the massive amount of data generated by IoE devices.
- *Cost*: Cost is another factor due to upfront costs such as infrastructure investment, system integration, and other investments. Companies always look for a return on investment before deploying the IoE solutions.
- *Standardization*: The protocols and other communication standards are the backbone of any network. The IoE networks are still in the development phase where the lack of protocols and standards leads to compatibility and integration issues. There is a need to establish new compatible standards and protocols for broader interoperability and adoption.
- *Legacy System Integration*: Integrating IoE networks with existing legacy systems is a challenging part due to a lack of necessary interfaces, standards, and communication protocols.
- *Regulatory Compliance*: There are several regulatory and compliance requirements for IoE network deployment. Data protection and privacy concerns must adhere to various regulatory and legal requirements. Organizations and companies need to ensure compliance with relevant rules and regulations to avoid any reputational consequences.
- *Energy Requirements*: The IoE devices are based on limited batteries and power resources. Energy optimization of these devices and finding new sustainable power resources are still a challenge, especially in inaccessible remote areas and environments.
- *Skill Gap*: Managing and implementing IoE networks need special skills and expertise. The scarcity of professionals and experts with IoE knowledge can pose a challenge for companies and organizations.

Addressing the discussed IoE network implementation and development challenges needs careful planning, expertise, and collaboration. Companies and organizations need to conduct the planning by using proper assessment, engagement of IT expert's vendors or consultants, and prioritizing security, scalability, and interoperability from the outset to ensure successful IoE implementation. These all issues and challenges need stakeholder attention through positive collaboration of industries, users, and government [6].

1.5 Acceptance and Sustainability

A digital twin is used to bridge advanced communication systems and computing technologies for data processes, information mirroring, and connected operations. There are three categories of digital twins including monitoring, simulation, and operations. The monitoring is utilized to mirror or virtual presentation of physical devices in IoE networks. The simulation of digital twins defined the software platform to predict and optimize the network performance. The operations are defined as the physical objects' transmission for optimal solutions to understand the physical objects' operations. Multi-access Edge Computing (MEC) is another emerging concept developed by the European Telecommunication Standards Institute (ETSI) as a distributed cloud architecture. The MEC is used as a backbone technology to transform traditional mobile networks and improve the IoE network's operations and services. Edge computing is closing to IoE end devices to improve the network delay and burden issues. This strategy also looks after the computational and caching processes at the edge like Base Station (BS) and Road Side Unit (RSU) and processes locally. The MEC also facilitates IoE devices with limited resources and low latency requirements [7].

The communication standards and protocols have a direct impact on IoE network services like data throughput, transmission delay, data delivery, and security. The IoE networks need resources for data communication. The various communication standards need optimal operations in virtual edge networks for better communication services. These requirements are important for better communication services including data rate, bandwidth, and security. The computation services are limited in IoE networks due to intensive tasks to support the new emerging applications and services. New technological-based architectures are the main focal point for companies for IoE network adoption. The IoE architecture is considering the networks and business requirements, enabling technology and application domains. The researchers proposed different types of technological architectures for IoE networks based on generic building blocks and solutions for IoE vision for better services to the users [8]. Technology-enabled business models are also one of the directions for the acceptance and sustainability of IoE networks. Digital technologies and relevant business development are the notions where people and things will overlap in a system. The major driver behind the IoE network's adoption and sustainability is the new emergence of globally interconnected networks. Interconnectivity of components in IoE networks is needed to support these networks for a computational resource on demand services.

The IoE networks are essential and able to cover several fields and offer cost-effective, fast, and interactive services. Smart city concept is also promoting the concept of IoE networks where all the networks and users are connected [9]. In recent years, quantum computing is another technology that will intersect with emerging communication technologies like 6G. The IoE networks are enabled with 6G technologies and offer ubiquitous distributed computing technologies and services. These networks are based on mixed reality, virtual reality, and extended reality applications for the formation of new services. The 6G mobile systems can support and fulfill the existing IoE network requirements for sustainability and adoption. Currently, the 5G technologies are merged with IoT networks where billions of devices are communicating for different purposes. However, with the new emerged concept of IoE, the existing technologies like 5G are not able to handle the new services and increasing demand for inter-cell interference. The 6G mobile systems will handle the massive converge of IoE networks by using integrated communication services such as AI-based wireless communication, advanced signal processing services, and full duplex and spectrum sharing services. The new IoE services will fulfill the industrial demand by offering computing intelligent systems. There is a need to improve the IoE network services for better data communication services, greater scalability, broader reach, and low-cost solutions [10]. The following key points need to be considered for IoE network's acceptance and sustainability:

- **Connectivity:** The connectivity requirement depends on reliable Internet infrastructure and widespread connectivity of devices and systems.
- **Interoperability:** This factor ensures the system working processes across different technologies and standards. The devices in IoE systems can communicate and share the data effectively, promoting acceptance and avoiding vendor lock-in.
- **Security and Privacy:** Security and user privacy are always the main concern of these networks due to the vast amount of data exchange among systems, devices, sensor nodes, and people. Encryption, authentication, trust, and privacy methods are essential for widespread acceptance and long-term network sustainability.
- **Standardization:** Standards and protocols are essential for sustainable network adoption and successful implementation of the IoE. Standards should cover the basic compatibility, interoperability, and scalability requirements for widespread adoption.
- **Scalability:** The scalability ensures the growth of the IoE expanding network and handles the increasing number of connected devices and the growing volume of data. This factor is significant for long time adoption of networks and their future growth.
- **Energy Efficiency:** Energy is always under consideration due to the limited resources of smart devices and complex network interconnections. Energy concern is crucial for the IoE network's sustainability and reducing environmental impact and ensuring long-term viability.
- **Data Governance**: Regulations and clear guidelines are required to govern the IoE networks in terms of data collection and storage. Transparent data governance strategies improve the user's trust and privacy and promote acceptance and sustainability.
- **Ethical Considerations:** Due to ethical concerns, this factor needs attention to ensure data ownership, user's consent, and potential biases in automated decision-making. These practices are important for fair network design and development phases for long-term sustainability.
- **Education and Awareness:** Due to the widespread acceptance of IoE networks, the education and awareness of users and other stakeholders are necessary to understand the network processes and to promote the network's services.
- **Economic Impact:** Economic values and benefit promotion are needed to enhance the values and acceptance of the network. Economic impact also refers to potential innovation, productivity gains, cost savings, and new business opportunities.

1.6 Proposed Five-Layer Conceptual Model

This section proposes a framework for IoE networks, which can be divided into five layers, namely, everything layer, communication layer, data layer, virtual layer, and application layer. The everything layer contains all devices involved in IoE networks like sensors, mobile devices, and other short- and long-range communication devices. The everything layer is further connected with the virtual layer where all things are managed and controlled by using simulations, optimization, and prediction. The virtual layer is also handling the service-related processes, hardware configuration, resource allocation, real-time operation status, and other monitoring tasks. The data layer handles the data generated from the everything and virtual layer. This layer is responsible to manage and analyze the collected data to refine it for edge or cloud computing storage. The communication layer handles all the processes related to edge and cloud computing. This layer also makes a connection between the other layer and working as infrastructure. All wired and wireless communication standards are handled in this layer. The four-layer conceptual model provides a useful way to understand the different components of an IoE system and how they work together. By understanding the different layers, it is possible to design and implement more effective and efficient IoE systems that can provide real value to users.

1.6.1 Everything Layer

This layer represents the physical devices or "everything" that is connected to the Internet, such as sensors, smart appliances, and wearables. The IoE network covers almost all areas and fields where the different devices, sensor nodes, and system are working with each other for data communications [11]. The devices of the IoE network are categorized based on their usage such as for medical healthcare, smart homes, vehicular networks, smart grids, and smart industrial devices. In healthcare wearable devices are included which are used to monitor several vital signs of patients and deployed inside and outside the patient's body. These devices are also used for normal heart monitoring, calorie recording, and normal routine monitoring. On the other hand, smart home appliances are equipped with smart sensor nodes to control the devices from smartphones by using Internet services. Smart metering and utility management applications are examples of smart grid systems to control and manage processes. Vehicle or intelligent transportation systems offered a plethora of applications where vehicles are moving and exchanging information in the network for safety and entertainment. The everything layer also covers the wireless and wired devices connected with the user's devices for further data processing. RF identification (RFID) tags and other low-power passive and active wireless devices are also used for smart industries. The scalability and reliability of these

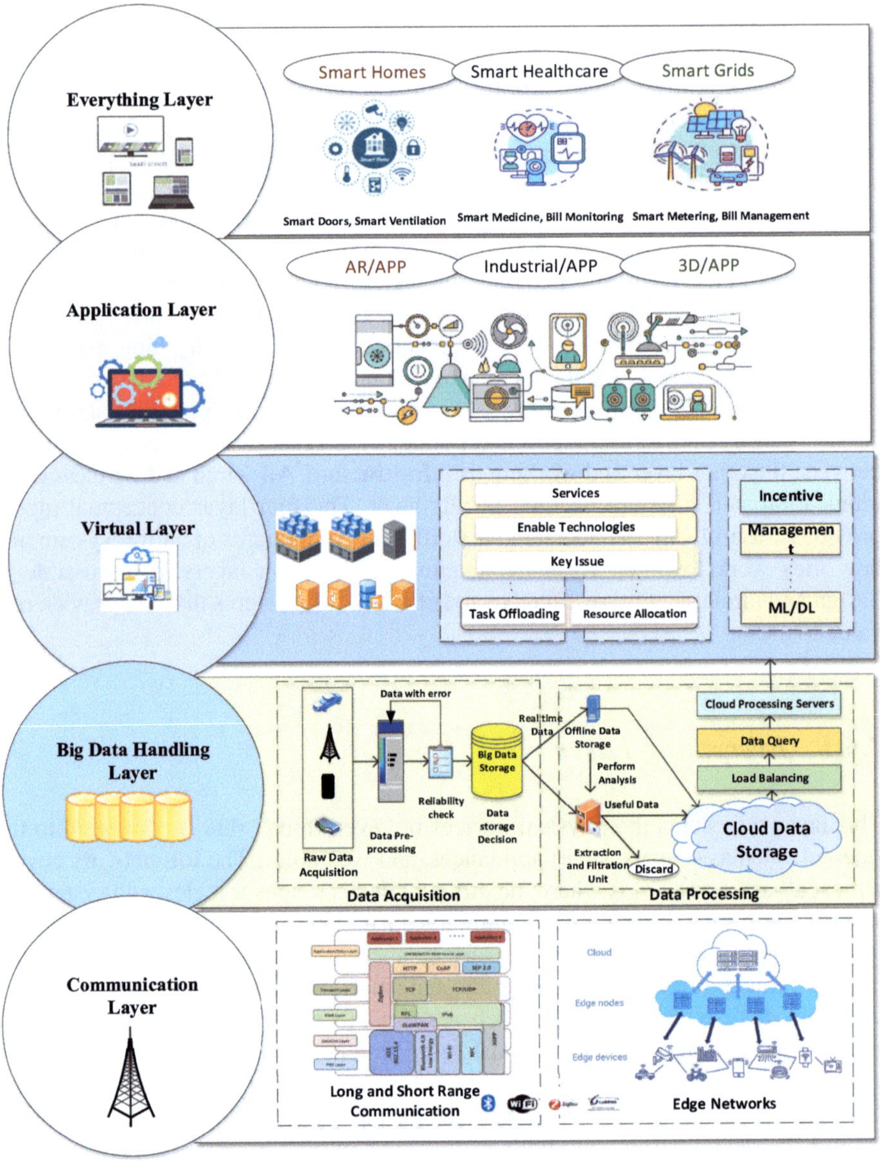

Fig. 1.2 Proposed five-layer model for IoE

interconnected different devices are still a challenge for IoE networks. Figure 1.2 shows the complete layer model structure with all entities.

The embedded and heterogeneous devices are separated in this layer where the low-power radio services are allocated by using well-established cellular network

Fig. 1.3 Example devices for IoE networks

services. The IoE devices are not only embedded devices but also included multi-vendor devices. Figure 1.3 shows the example devices of IoE networks.

1.6.2 Application Layer

This layer represents the applications and services that use the data generated by the everything layer to provide value to users. This can include smart home applications, healthcare applications, and industrial automation applications. The applications and services need attention in terms of the computational capabilities of IoE devices and their storage space. Due to the limited battery life of IoE devices, the complex applications degraded the services, and users faced the delay, overhead, and battery depletion issues. Robotics also uses AI applications which need more resources as compared to smartphones and home appliance devices. This layer provides common services running on everything layer. This layer also offers interoperability by selecting better-operating systems to support the IoE network solutions. Programming languages are playing a crucial role to design applications for IoE network devices that are capable to handle network discovery, group communication processes, and network connections and support the context-aware services and scalability. There are different studies conducted for IoT network application requirements and presented the conceptual view based on service orientations [12].

1.6.3 Virtual Layer

The virtual layer provides programming support to other layers for user application execution in IoE networks. This virtual-based layer divided the applications into separate modules and further linked them with the application layer. This layer holds the virtual machines placed between the operating systems and applications or at the system level. The concept of a virtual layer or virtual machine-based middleware was started in 2012 [13] where authors initiated the project *Mate* which introduced resources management by allowing the update to the virtual machines' middleware. There are some other examples of virtual Java-based solutions designed for IoT networks like MagnetOS and SensorWare [14, 15]. However, these solutions were designed for static nodes and neglected the mobility of users. The IoE network's vision is to provide connections to everything by using a common infrastructure and handling complex processes. The proposed virtual layer provides the agent-based solution which reduces the complexity of systems by using the higher-level policies.

1.6.4 Data Handling Layer

This layer represents the data generated by the everything layer, which is collected, stored, and analyzed. This data can include sensor data, user data, and other forms of data. The raw data is generated from IoE devices from the everything layer and further transmitted to the communication layer for updating and modeling and decision-making. The big data handling layer classified the data into its volume, velocity, and variety. This layer contains the data handling tools for data processing including Apache Spark, Apache Flink, and Apache HBase [16]. This layer is responsible for data acquisition, retention, transport, processing, and data leverage. In the acquisition part, the data is collected, and clean processes are applied before being transmitted to another layer. Data retention refers to the policies required to manage the data such as privacy and legal concerns and encryption methods. The big data layer also handles the load balance, replication, and continuity processes. The IoE network data is based on real-time analytics and based on rules and policies which define the rules for data input [17]. Data analytics is also performed in terms of descriptive, predictive, and perspective analytics. Descriptive means what happens and what is happening, whereas predictive means what will happen and why happen. The prescriptive means what should I do and why should I do it. These processes will help for new business opportunities, predict the future, and provide better decision-making strategies.

1.6.5 Communication Layer

This layer represents the network infrastructure that enables the communication between the everything layer and the data layer. This includes wired and wireless networks, protocols, and standards. The large scale of data transmission is needed to achieve communication among IoE devices and other edge and cloud-based backbone devices [18]. There are different types of communication involved in this layer including device-to-device communication, device-to-infrastructure communication, and infrastructure-to-infrastructure communication. The device-to-device communication is the type where devices or everything is communicating with each other by using a short- or long-range communication standard [19]. The device-to-infrastructure communication is achieved when the devices are communicating with an edge or cloud-based services by using wired or wireless communication technologies. The last type of infrastructure-to-infrastructure communication is achieved when the backbone devices further communicate with cloud and edge devices and are further connected with satellite communications and other Internet platforms. Due to different communication standards and protocols, all types of communication need different requirements in terms of bandwidth, latency, and capacity. Table 1.1 shows the used communication technologies in IoE.

Table. 1.1 IoE technologies and details

S/No	Technology	Description	Example use cases
1	RFID (radio frequency identification)	Uses electromagnetic fields to automatically identify And track tags attached to IoE objects	Used for supply chain tracking, inventory management, and access control systems
2	Zigbee	Short- and long-range low-power wireless communication protocol for low-data-rate applications	Used for industrial automation, smart homes, smart grid systems
3	Bluetooth	Short-range wireless technology for transmitting data between devices	Used for smart healthcare systems, smart home devices, and other wearable devices
4	Wi-fi	Wireless networking technology for local networks that enables devices to connect to the internet	Used for smart healthcare, smart industries, systems, and smart home devices
5	LTE-M (long-term Evolution for machines)	Cellular wide coverage and low-power technology optimized for IoE devices	Asset tracking, smart cities, utility metering
6	LoRa (long range)	Low-power, wide area network (LP-WAN) technology for long-range communication	Smart agriculture, environmental monitoring, remote sensor networks
7	NB-IoE (narrowband IoE)	Cellular network technology designed for low-power, low-cost IoE devices	Smart meters, asset tracking, logistics, and transportation monitoring
8	MQTT (message queuing telemetry transport)	Lightweight messaging protocol for efficient communication between devices	Remote monitoring, telemetry, IoE data collection and analysis
9	CoAP (constrained application protocol)	Lightweight application-layer protocol for constrained devices and networks	Smart energy systems, home automation, smart cities
10	Edge computing	Decentralized computing infrastructure that brings processing and analytics closer to IoE devices	Real-time analytics, low-latency applications, reducing cloud dependency

1.7 Discussion and Findings

The IoE networks represent a significant shift in the way where the users are connected by using different smart devices. By connecting everyday objects to the Internet, IoE has the potential to revolutionize many areas of our lives, from healthcare and transportation to agriculture and manufacturing. However, it also raises important questions about privacy, security, and the ethical implications of such a connected world. One of the key benefits of IoE is the ability to gather and analyze large amounts of data in real time. This data can be used to improve efficiency, optimize processes, and provide new insights and value to users. For example, in healthcare, IoE can be used to monitor patients' health in real time, alerting doctors and caregivers to any issues before they become serious. In manufacturing, IoE can be used to improve efficiency and reduce downtime in factories, allowing businesses to produce goods more quickly and with fewer errors.

However, IoE also raises important concerns about privacy and security. With so many devices connected to the Internet, there is a greater risk of cyberattacks and data breaches. Ensuring the security and privacy of IoE devices and data is critical to ensuring that users can trust and rely on IoE systems. In addition, IoE raises important ethical and social implications. For example, IoE could potentially exacerbate existing inequalities, as those who cannot afford to invest in IoE systems may be left behind. Ensuring that IoE is implemented in a safe, secure, and responsible manner, with consideration for its social and ethical implications, is critical to realizing the full potential of this technology.

Overall, the IoE represents a significant opportunity to improve our lives and our world through greater connectivity and intelligence. However, we must approach this technology with caution, ensuring that we address the challenges and issues that arise as we move toward a more connected world.

1.8 Conclusion

The IoE is one of the prominent fields where almost everything is connected to the Internet for monitoring and data communication services. These networks are based on high-end technologies and new standards to fulfill the existing communication demand. Due to complex and heterogeneous systems, these networks have suffered from scalability, reliability, security, communication, and data analytic issues. This chapter presented the basic operation of IoE networks, applications, and used technologies. This chapter also presented a five-layer framework including everything layer, application layer, virtual layer, data layer, and communication layer. The proposed framework will address the existing issues and challenges to fulfill the network requirements. In the future, the proposed model will integrate with other networks like satellite communication, drone technologies, and robotics.

References

1. Liu Y, Dai H-N, Wang Q, Shukla MK, Imran M (2020) Unmanned aerial vehicle for internet of everything: Opportunities and challenges. Computer communications 155:66-83. https://doi.org/10.1016/j.comcom.2020.03.017
2. Qureshi KN, Jeon G, Piccialli F (2020) Anomaly Detection and Trust Authority in Artificial Intelligence and Cloud Computing. Computer Networks:107647. https://doi.org/10.1016/j.comnet.2020.107647
3. Butt N, Shahid A, Qureshi KN, Haider S, Ibrahim AO, Binzagr F, Arshad N (2022) Intelligent Deep Learning for Anomaly-Based Intrusion Detection in IoT Smart Home Networks. Mathematics 10 (23):4598. https://doi.org/10.3390/math10234598
4. Ali Z, Qureshi KN, Al-Shamayleh AS, Akhunzada A, Raza A, Butt MFU (2023) Delay Optimization in LoRaWAN by Employing Adaptive Scheduling Algorithm with Unsupervised Learning. IEEE Access:1–1. doi:https://doi.org/10.1109/ACCESS.2023.3234188
5. Aliero MS, Qureshi KN, Pasha MF, Ghani I, Yauri RA (2021) Systematic Mapping Study on Energy Optimization Solutions in Smart Building Structure: Opportunities and Challenges. Wireless Personal Communications:1-37. https://doi.org/10.1007/s11277-021-08316-3
6. Antonios P, Konstantinos K, Christos G (2023) A systematic review on semantic interoperability in the IoE-enabled smart cities. Internet of Things:100754. https://doi.org/10.1016/j.iot.2023.100754
7. Iannacci J (2018) Internet of things (IoT); internet of everything (IoE); tactile internet; 5G–A (not so evanescent) unifying vision empowered by EH-MEMS (energy harvesting MEMS) and RF-MEMS (radio frequency MEMS). Sensors and actuators a: physical 272:187-198. https://doi.org/10.1016/j.sna.2018.01.038
8. Langley DJ, van Doorn J, Ng IC, Stieglitz S, Lazovik A, Boonstra A (2021) The Internet of Everything: Smart things and their impact on business models. Journal of Business Research 122:853-863. https://doi.org/10.1016/j.jbusres.2019.12.035
9. Padhi PK, Charrua-Santos F (2021) 6G enabled industrial internet of everything: Towards a theoretical framework. Applied System Innovation 4 (1):11. doi: https://doi.org/https://doi.org/10.3390/asi4010011
10. Anand S, Ramesh MV Multi-layer architecture and routing for internet of everything (ioe) in smart cities. In: 2021 sixth international conference on wireless communications, signal processing and networking (WiSPNET), 2021. IEEE, 411–416. https://doi.org/10.1109/WiSPNET51692.2021.9419428
11. Kruger CP, Hancke GP Benchmarking Internet of things devices. In: 2014 12th IEEE International Conference on Industrial Informatics (INDIN), 2014. IEEE, pp 611–616. https://doi.org/10.1109/INDIN.2014.6945583
12. Yaqoob I, Ahmed E, Hashem IAT, Ahmed AIA, Gani A, Imran M, Guizani M (2017) Internet of things architecture: Recent advances, taxonomy, requirements, and open challenges. IEEE wireless communications 24 (3):10-16. doi:https://doi.org/10.1109/MWC.2017.1600421
13. Levis P, Culler D (2002) Maté: A tiny virtual machine for sensor networks. ACM Sigplan Notices 37 (10):85-95. https://doi.org/10.1145/605432.605407
14. Kirsch CM, Sanvido MA, Henzinger TA A programmable microkernel for real-time systems. In: Proceedings of the 1st ACM/USENIX international conference on Virtual execution environments, 2005. pp 35-45. doi:https://doi.org/10.1145/1064979.1064986
15. Boulis A, Han C-C, Shea R, Srivastava MB (2007) SensorWare: Programming sensor networks beyond code update and querying. Pervasive and mobile computing 3 (4):386-412. https://doi.org/10.1016/j.pmcj.2007.04.007
16. Ahmed E, Yaqoob I, Hashem IAT, Khan I, Ahmed AIA, Imran M, Vasilakos AV (2017) The role of big data analytics in Internet of Things. Computer Networks 129:459-471. https://doi.org/10.1016/j.comnet.2017.06.013

17. Qureshi KN, Alhudhaif A, Arshad N, Kalsoom U, Jeon G (2021) Data analysis based dynamic prediction model for public security in internet of multimedia things networks. Multimedia Tools and Applications. doi:https://doi.org/10.1007/s11042-021-11462-2
18. Qureshi KN, Alhudhaif A, Haider SW, Majeed S, Jeon G (2022) Secure Data Communication for Wireless Mobile Nodes in Intelligent Transportation Systems. Microprocessors and Microsystems:104501. https://doi.org/10.1016/j.micpro.2022.104501
19. Qureshi KN, Bashir F, Abdullah AH (2019) Distance and signal quality aware next hop selection routing protocol for vehicular ad hoc networks. Neural Computing Applications:1-14. https://doi.org/10.1007/s00521-019-04320-8

Chapter 2
Cybersecurity Threats and Attacks in IoE Networks

Bahareh Pahlevanzadeh and Sima Ahmadpour

2.1 Introduction

In the age of Industry 4.0, the Internet of Everything (IoE) is one of the prominent technologies for data communication. In fact, the IoE goes beyond the traditional Internet of Things (IoT), which focuses primarily on connecting devices by connecting people and processes in more meaningful ways. The IoE is the culmination of several long-term technological developments. The IoE is the result of the widespread adoption of novel technologies and paradigms such as cloud computing, edge computing, and 5G/6G networks, to create a more connected and intelligent environment, the proliferation of devices and sensors, and the availability of affordable and powerful computing resources.

This chapter discusses the IoE and a comprehensive view of the potential cybersecurity threats and attacks in IoE networks and finally proposes an IoE 3D cybersecurity model based on three dimensions of the multi-layered architecture of IoE, including components and entities (first dimension), IoE and its multi-layered ICT architecture (second dimension), and IoE and its enabling technologies (third dimension), in the context of digital transformation. To provide a better overview, this chapter presents an overview of IoE and its major components and technologies (in Sect. 2.2), various application areas of IoE (in Sect. 2.3), the IoE threat landscape and threat modeling (in Sect. 2.4), and, finally, various cybersecurity threats,

B. Pahlevanzadeh
School of Informatics and Cybersecurity, Technological University Dublin (TU Dublin), Dublin, Ireland
e-mail: Bahareh.pahlevanzadeh@tudublin.ie

S. Ahmadpour (✉)
Graduate School of Business, Universiti Sains Malaysia (USM), Penang, Malaysia
e-mail: ahmadpour.sima@usm.my

K. Naseer Qureshi et al. (eds.), *Cybersecurity Vigilance and Security Engineering of Internet of Everything*, Internet of Things,
https://doi.org/10.1007/978-3-031-45162-1_2

security considerations, solutions, and countermeasures considering various related works in this area and a proposed IoE 3D cybersecurity model (in Sect. 2.5), and finally, we ended the chapter with a summary and a conclusion.

2.2 Internet of Everything

In 2013, Cisco introduced the term "Internet of Everything" (IoE), which refers to the interconnectedness of people, processes, data, and things through networked connections. The IoE aims to create a more valuable and relevant network of connections than ever before, turning information into actionable insights and generating new capabilities. This concept builds upon the Internet of Things (IoT) by including not just physical devices but also people and processes and seeks to create a seamless and intelligent ecosystem that can generate new efficiencies and economic value [1]. The IoE has significant implications for a wide range of industries, including healthcare, transportation, manufacturing, and smart cities, and has the potential to transform the way that businesses and organizations operate.

2.2.1 IoE Components

There are four components/pillars of IoE as follows:

People: Connect people around the world with various valuable and significant resources. Various devices in and on people (PCs, smartphones, tablets, etc.) are constantly connected to the Internet, helping to establish more valuable and reliable communication between other people and devices.

Everything: In IoE networks, the devices are connected for data communication and processes. The proliferation of intelligent things has driven technological advancements toward intelligence-based edge computing [2]. Every single devices is connected by using other communication systems for data processing, collection, and storage [3]. In this scenario resource-intensive and complex security solutions support the traditional networks. Conversely, IoE systems necessitate security protocols that exhibit lower resource consumption while maintaining a delicate equilibrium between resource consumption and security.

Data: Using data to make better and more responsible decisions. Data can be collected from anywhere for decision-makings in various aspects of our daily lives. Processing and using the collected data also play an important role in the case of IoE. Machine learning and data analytics are anticipated to lead contextualized data processing in big data and IoE analysis. Each linked user and process generate data, which must be effectively gathered, processed, summarized, categorized, and evaluated.

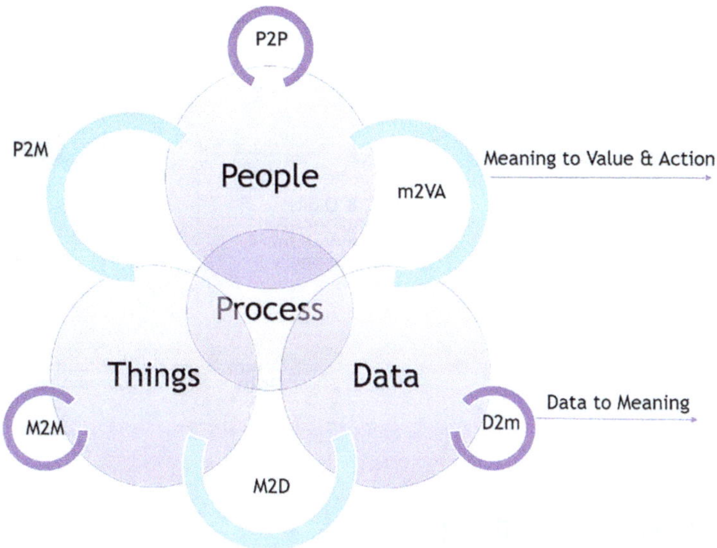

Fig. 2.1 IoE components

Process: Transmitting the right data to the right person or place node/machine) at the right time [4]. Unlike the typical application-specific traditional processes, the processes in the IoE represent the more substantial resource in furnishing a ubiquitous user experience.

IoE dramatically expands the scope of IoT by adding components that can deliver an even better experience to businesses, individuals, and countries. Rather than just relying on things to interact with their environment, IoE can leverage all of the associated data and processes to make the IoT more relevant and valuable to people. As vendors and scientists have noted, the ultimate goal of IoE is to deliver these benefits while maintaining a strong focus on operational efficiency, security, and privacy. By doing so, IoE can help to create a more intelligent, interconnected, and sustainable world, in which technology is used to improve the lives of people and communities around the globe. Figure 2.1 shows the IoE components.

2.2.2 IoE Key Enabling Technologies

The IoE ecosystem is built upon several key technologies, including cloud computing, big data analytics, 5/6 G networking, artificial intelligence (AI), and the IoT, which are known as enablers in digital transformation (Industry 4.0). These technologies enable devices to connect to the Internet and transmit data that can be analyzed and processed in real time, resulting in valuable insights and actionable intelligence. Figure 2.2 shows the IoE and digital transformation enablers.

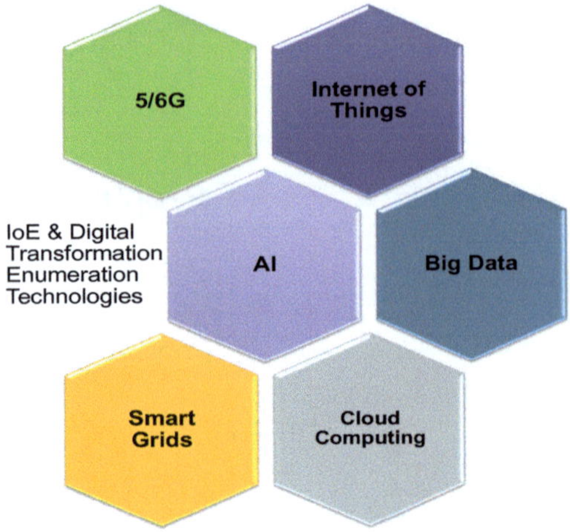

Fig. 2.2 IoE and digital transformation enablers

2.3 IoE Threat Landscape

The IoE threat landscape refers to the potential cybersecurity risks and threats posed by the interconnectedness of people, processes, data, and devices in the digital ecosystem. As IoE technologies proliferate, the risk of cyberattacks also increases, as cybercriminals can target any point in the IoE ecosystem to gain unauthorized access or disrupt critical services. In general, understanding the current IoE threat landscape is critical to maintaining the security and integrity of IoE devices, networks, and systems, as well as protecting sensitive data and assets. The IoE threat landscape encompasses a wide range of potential attacks. This section presents the landscape of IoE security issues and threats based on the four main pillars of IoE.

2.3.1 IoE Threat Modeling

The threats of data security hold immense significance in the current century because even a minor information breach can have grave implications for an organization's reputation or finances. Over the last two decades, people's daily lives have become increasingly connected with the Internet and digital world [5]. The necessity of protecting information is undeniable, and any system that is developed and in operation must possess sufficient protection against the existing threats. While it is vital to have comprehensive information protection, it is equally important to consider the potential threats that are exclusive to a particular information system. By advancing technology, attackers can access valuable data easier than before; therefore information system security becomse a critical concern. Threat modeling has played an important role in the integration of security into software systems. It enables us to identify crucial areas of the design that require to be protected. Over time, several approaches and methodologies for threat modeling have been developed and are employed in the development of secure web applications, API applications, and the Web of Things (WoT) as one of the components of the IoE.

Furthermore, in order to ensure the secure deployment of IoE, it is imperative to take into account a variety of mechanisms and parameters based on four key pillars: people, process, data, and things. Given that IoE technology encompasses a wide range of devices, spanning from small compact embedded processing chips to expansive high-end servers, a multi-layered security architecture must be implemented to tackle security concerns across different levels. Several security architectures have been proposed for IoT environments, which can generally be classified into three categories: three-layer, four-layer, and five-layer architectures [6]. As IoE represents the next stage of IoT evolution, the existing IoT layered architecture comprises four primary layers (ITU-T has proposed), namely, the perception layer, network layer, support layer, and application layer.

2.4 IoE Cybersecurity Architecture

There are several factors that must be considered in security in general. Like any other computing system, IoE devices are vulnerable to cyberattacks. IoE has indeed brought a new set of vulnerabilities, risks, and challenges in various areas. Limited computation capabilities, hardcoded passwords, insecure networks, insecure employees, lack of secure updates, massive data transfer, and limited computation capabilities are some of the vulnerabilities that can exist within IoE systems, and they can expose interconnected devices and systems to security risks [6].

The risks such as lack of awareness, limited control, inadequate monitoring and governance, immature response mechanisms, lack of resources, and lack of comprehensive security solutions can all pose significant threats in the context of IoE such as denial of service (DoS), malware, ransomware, social engineering, etc. that would be presented in the following sections in more details. While IoE is driving performance and modernity in almost all sectors through increased connectivity and automation, it is also increasing new cybersecurity risks and vulnerabilities to the security of data, systems, and people; therefore, the frequency of cyberattacks and the number of security breaches continue to grow. Securing IoE networks involves securing their various components [7, 8]. In order to ensure a secure deployment of the IoE, it is imperative to consider a variety of mechanisms and parameters [9]. Figure 2.3 shows the considered features for the IoE environment.

A smooth and effective integration of diverse connected devices, processes, people, and data is supported by the IoE. In addition, Fig. 2.4 shows the multi-layered architecture of IoE and its end-to-end security including three dimensions of four pillars (dimension 1), OSI-based elements and layers (dimension 2), as well as digital transformation enablers (dimension 3) which are as follows.

Fig. 2.3 Considered features for the IoE environment

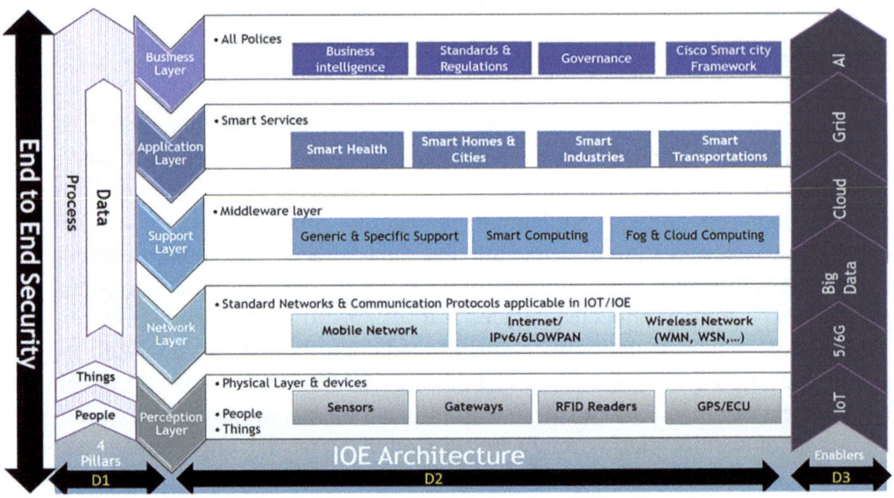

Fig. 2.4 IoE multi-layered architecture (3D cybersecurity model: pillars, elements, enablers)

2.4.1 Cybersecurity Vulnerabilities, Threats, and Attacks in Perception Layer

The "things" in IoE refer to the process of connecting physical devices and objects to the network, such as sensors, wearables, and appliances. These things have the ability to sense, process, and transmit data. The perception layer, which is also commonly referred to as the sensor, hardware, or edge layer, is a fundamental component of the IoE architecture. The perception layer is comprised of two distinct elements, namely, the perception channels or nodes (including controllers, sensors, and other related components) and the perceptual networks that are connected to the network layer [10].

Within the perception layer, the perception nodes or physical devices gather a wide range of information, such as RFID readers, GPS, smart meters, ECU (electronic control unit), and smart home electronics, among others. These sensors and devices gather data regarding object attributes and environmental conditions, to create intelligent data that can transform and transmit instructions to the perception network layer. At the IoE perception layer, securing equipment through the deployment of public-key encryption and high-frequency communication programs are impossible due to limited storage and computing power. Networked devices are continuously exposed to various threats, which are increasing very fast. The conventional security measures are inadequate for low-power devices to satisfy the security requirements.

2.4.2 *Identity-Based Trust and Privacy Provocation*

Attackers use the collected data of versatile user categories to carry out a various range of malicious activities, such as using it for hostile purposes and linking identity to a specific individual.

- *Location-based tracking:* Attackers aim to compromise user privacy and trust by trying to trace the location of users through mobile and wearable devices and extracting sensitive information by taking advantage of user history.
- *User profiling:* Attackers execute user information to construct user profiles, which may furnish users with additional information that may not be interesting or disclose. The aforementioned behavior constitutes a breach of trust as well as confidentiality.

IoE devices could maliciously collect private user data about microphones, cameras, keystrokes, and proximity to the user. This information could leak out and be shared with undesirables. Users are maliciously observed asking various queries and analyzing the user's underlying response.

2.4.3 *Spoofing and Sybil Attacks*

Spoofing attacks refer to malicious activities in which an attacker disguises or impersonates themselves as someone or something else to deceive or gain unauthorized access to a system [11, 12]. Subsequently, the attackers manage to acquire entry into another device or user present within the network. Spoofing attacks can be categorized into two distinct types: (i) link-layer spoofing, where all communication between two parties is subjected to spoofing, and (ii) end-to-end layer spoofing, where specific services can be spoofed by the attacker [13]. Attackers have the ability to initiate various types of spoofing attacks. The objective of this attack is to undermine authority or power within a trusted system by attaining a majority of influence in the network.

2.4.4 Access-Level Attacks

Generally, there exist two types of attacks in this category:

Active attacks: In this case, the attacker intends to create disruption among authorized nodes by impersonating the identity of another node or manipulating routing information.

Passive attacks: Passive attacks are mostly the intruder eavesdrops on communications between legitimate senders and receivers to obtain the transmitted data.

2.4.5 Transmitting Data Attacks

IoE devices are capable of detecting and gathering data regarding their surrounding environment. However, it is noteworthy that attackers may exploit communication channels that were not intended to be utilized between existing device peripherals. By doing so, these attackers are able to manipulate critical sensor configurations [14].

- *Transmission via light sensors*: The technology for compromising signals and transmitting malicious signals involves the use of light sensors, which facilitate the transmission of data packets through turning a light source on and off.
- *Transmission via magnetic sensors*: Sensors are compromised when the magnetic fields of peripheral devices change. An attacker can spoof magnetic sensor data by manipulating the atmospheric magnetic field surrounding the device; an attacker can falsify magnetic sensor data by spoofing.
- *Transmission via audio sensors*: The activation of malware can also occur by utilizing the functionality of audio sensors. Microphones are capable of detecting audio signals at frequencies that are significantly lower than the audible range, and this type of audio can be used to send triggering messages to bypass device security measures.

2.4.6 Attacks Based on Device Property

The devices are categorized into low- and high-end devices where each attack behaves differently. High-end device attacks disturb the processes to connect to the IoE network and launch attacks from any location and at any time [15, 16]. In low-end device class attacks, the devices used for this attack can connect between the system and the outside world via wireless connections.

2.4.7 Attacks Based on Adversary Location

At any given time and from any location, an intruder can launch an attack on an IoE system. This includes both internal and external individuals who may initiate such attacks [17]. The primary objective of most external attacks is to steal users' private data through the utilization of malware like worms, Trojan viruses, phishing attempts, and similar techniques. By exploiting these methods, a hacker can establish unauthorized control over an IoE device situated at any location within the IoE network.

- *Internal attacks:* An insider threat refers to an individual who possesses legitimate access to a network or system and intentionally misuses data or exploits their authorized access to engage in wrongful activities.

2.4.8 Attacks Based on Attack Strategy

To initiate an attack, the attacker adopts a strategic approach to insert and execute the malicious code they designed into an IoE device to disrupt the IoE network.

- *Physical attacks:* The attacker must have direct physical or network access to the infrastructure of the IoE network. By altering the instructions or structure of the system, the attacker has the capability to disrupt either a partial or full IoE network.
- *Logical attacks:* Physical access is unnecessary, and the attackers do not cause physical harm to the device; instead, they render the communication channel unable to function [18].

2.4.9 Insecure Initialization and Configuration

Insecure initialization and configuration occur when devices, systems, or applications are not properly set up or configured during initial installation or deployment. These vulnerabilities can leave the system exposed to potential attacks or unauthorized access.

2.5 Cybersecurity Considerations and Solutions in Perception Layer

Existing solutions to the jamming threat include measuring signal strength, calculating packet transmission rates, error-correcting codes, and changing frequencies and locations. The key solutions to identifying privacy and trust are anonymization techniques and manage the local computing capabilities. In location-based tracking, the user must be aware of the nature of the information and to control and block such information, virtualization technologies. The interactions between privacy and data protection can be resolved by ISO/IEC Guidelines for Cybersecurity standards and GDPR (General Data Protection Regulation). Signal strength measurements and channel estimates cover Sybil and spoofing attacks. Insecure initialization and configuration are maintained by setting data transfer rates between nodes and introducing artificial noise. Avoidance of software/firmware access to USB, hardware-based TPM modules, and avoidance of test/debugging tools are suggested to overcome insecure physical interfaces. The multi-layered intrusion detection system is proposed to prevent sleep deprivation attacks. Phishing secret information is solved by several following solutions including potential employee identification strategy, psychological well-being of individuals, scalable security mechanism, graph-based strategy, benchmarking framework, semi-automatic feature generation-based machine learning strategy, potential employee identification strategy, psychological well-being of individuals, scalable security mechanism, graph-based strategy, benchmarking framework, semi-automatic feature generation-based machine learning strategy, a scalable security mechanism, graph-based strategy, benchmarking framework, and semi-automatic feature generation-based machine learning strategy which are all other proposed solutions and countermeasures used in various types of research to overcome the perception layer threats in IoE.

2.5.1 Cybersecurity Vulnerabilities, Threats, and Attacks in Network Layer

The IoE is centered around the network layer, which utilizes various communication protocols such as Wi-Fi, Bluetooth, and Zigbee to ensure data transmission between IoT/IoE components. This layer facilitates the data transmission and facilitates the different networks. Within the Internet architecture, security measures are primarily designed to address human behavior and may not seamlessly extend to the security mechanisms required between machines within the IoT/IoE. This is due to the diverse network infrastructures that devices utilize to communicate with one another.

The proliferation of IoT/IoE devices necessitates an examination of the security implications of common authentication methods. The confidentiality, accuracy, and comprehensiveness of data must receive special attention at this layer, and measures should be taken to address potential DDoS attacks and risks [19]. To mitigate these risks, Next Generation Networks (NGN) have become increasingly prevalent, with IPv6-based network security mechanisms being explored as a more secure alternative to IPv4. Information gathering technology and social engineering remain a concern, as hackers can collect vast amounts of private information. To address these issues, security solutions should be implemented. Mobile networks are also susceptible to attacks, such as DoS tracking, bluesnarfing, bluejacking, bluebugging alteration, corruption, and deletion [20]. Within this layer, two primary attacks are of concern: routing attacks and replay attacks.

- *Replay Attacks*: An attacker tries to transmit a packet that had been previously received by the intended recipient in order to gain system trust. This type of attack is employed to authenticate and destroy certificates.
- *Routing Attacks*: Through the act of forging or retransmitting routing data, an attacker with malicious intent can intentionally create loops in the routing process. This is done to resist the transmission of data, manipulate the length of the path, provoke error messages, prolong network latency, or intercept and divert network traffic. Additional types of routing attacks are outlined below [21].
- *Sinkhole or Blackhole Attack,* where attacker asserts that they possess a high-quality path, thereby granting them the ability to pass any packet through the fake path.
- *Selective Transformation*: Attackers can either select specific packets or completely discard them.
- *Wormhole Attack*: The attacker could register packets at a specific location within the network, relocate them to another terminal, and subsequently transmit them into the network. As a consequence, this activity disrupts the network's functionality and significantly impacts the performance of routing performance [22].
- *Sybil Attack*: Thus, this grants the attacker the ability to have multiple identities in relation to other objects within the network.

Suggestions and solutions for enhancing security at the network layer can be divided into three main categories: physical networks, remote connections, and wireless networks as discussed in the next subsections.

2.5.1.1 Security Solutions for Wired Networks

Implementing physical security measures within the network involves several practices, such as deploying closed-circuit television (CCTV) cameras, employing entry card systems to track individuals entering the premises, and establishing secure areas to deter unauthorized access.

Incorporating security mechanisms such as firewalls, intrusion prevention/detection systems (IPS/IDS), and access control lists (ACLs). These measures help protect against unauthorized access and potential threats.

2.5.1.2 Security Solutions for Mobile Networks

Implementing robust authentication methods like multi-factor authentication (MFA). This is an additional layer of security to authenticate authorized users seeking remote network access.

Facilitating secure communication channels for employees accessing the organization's network, it is recommended to utilize reliable communication channels like site-to-site VPN (virtual private network). This ensures a secure and encrypted connection between remote locations and the organization's network.

2.5.1.3 Security Solutions for Wireless Networks

Utilization of secure IoE device configurations when connecting wirelessly.

Utilization of cryptographic algorithms and authentication.

2.5.2 Cybersecurity Vulnerabilities, Threats, and Attacks in Support/Middleware Layer

This layer is meant to offer an application layer support platform that is trustworthy and dependable. For gathering, processing, and interpreting the data gathered by the devices, this layer is essential. The fundamental concept is that, based on the analysis of patterns in the gathered data and the categorization of this information for various use cases, the entire digital environment may be provided for IoE. The present layer is host to an array of diverse forms of intelligent computing that are systematically organized through the medium of cloud computing and grids [13]. The layer plays a critical role in effecting mass data processing and rationalizing intelligent network behavior decisions. It is a pressing concern to enhance the efficacy of malicious detection by utilization of powerful encryption algorithms and protocols to enhance effectiveness [1].

This layer contains a number of security threats and weaknesses, such as the management identity and the heterogeneity of IoT devices; this could prevent the transfer of information to a valid node. The complexity of the system, physical security, encryption, infrastructure security, user identification, security management practices, improper software configurations [20], privacy risks, and other related factors pose additional risks to the data access control within this layer. There are two DoS attacks situated within the support layer, which is interconnected and insider malicious attack that involves an authorized user accessing the data of other users, posing a highly intricate threat. It necessitates the implementation of various mechanisms to mitigate the risk.

2.5.3 Cybersecurity Considerations and Solutions in Support/ Middleware Layer

NoSQL authentication offers outstanding deployment of networks as a viable approach and promising solution that can be implemented in the IoE paradigm. Access controls can be put into place using NoSQL by managing databases for authenticated devices. Data leakage, information theft, tampering, and rejection are all threats from a wide spectrum of attacks. To successfully prevent data tampering and guarantee data privacy and confidentiality, data exchanges must be encrypted. Because IoE has resource limitations, data security methods must be resource-efficient to function well. There is currently no implementation of RSA algorithms for IoE, despite the fact that they have been utilized for data security in IoT. Data protection would be enhanced, and the system's ability to handle data leaks would be made possible by RSA and authentication methods.

In the IoT, hash algorithms have been actively used to examine data integrity as it is being transmitted across various nodes. Encryption could be employed to enforce security brought on by side channels, sniffing, and interception attacks. Additionally, given domain-specific features, such as available resources, transaction frequency, data rate, and target utilization of data, encryption could be used from an IoE standpoint. However, it is important to match the resource needs and processing capabilities of the target IoE device as closely as possible when using security solutions for data. Creating shared cryptography for communication is one of the effective methods, which lowers the IoT gateway's overhead. Compared to other cryptography techniques, it uses fewer network resources overall and has lower latency.

Although it uses more power, its performance is somewhat worse than that of the symmetry key and public key cryptography methods. The IoE paradigm still requires the development of novel, resource-efficient solutions that improve transaction security. Recent studies propose that hybrid encryption approaches ensure both feasible resource use and higher security. Additionally, in order to prevent configuration problems, the communication devices must use the same cryptographic suites. Using standardized cryptographic algorithms is a reliable way to prevent configuration problems with data security in IoE. Multi-factor cryptographic systems will offer a viable solution for the IoE's wide variety of devices and large-scale networks.

In the IoE paradigm, digital signatures offer an appropriate technique to guarantee the security and privacy of data among various levels and end devices. Contrary to popular belief, these methods are more effective than RSA and need less processing power than AES. Digital signatures, however, have constraints that are domain-specific since IoE devices may employ various routing protocols. Because IoE traffic originates from numerous interconnected data sources and adversaries occasionally transmit malicious packets to analyze network configurations, traffic filtering algorithms offer an effective defense against attacks. This method gets over the restrictions imposed by the various platforms.

The key security factors are discussed as follows:

Deploy security solutions within virtual machines, including regular operating system updates, defining appropriate access protocols for virtual machines (VMs), and implementing robust control mechanisms within the applications.
Secure the data stored in cloud environments by employing suitable technologies and authorized encryption algorithms.
Developing solutions for crisis recovery and ensuring service continuity involve implementing measures such as creating VM snapshots, performing regular backups, and having standby VMs available at the cloud provider's site.
Protect web applications by employing host-based firewalls to detect and prevent malicious traffic, alongside the utilization of intrusion prevention/detection systems (IPS/IDS).
Implement comprehensive log monitoring, particularly for authorized users, and effectively manage event logs from multiple sources using SIEM (security information and event management) solutions to analyze security incidents.
Utilize authentication schemes such as key exchange, credential systems, identity authentication, and IACAC (capability-based access control).
NoSQL authentication RSA combined with authentication techniques and hash mechanisms.
Employ domain-specific encryption techniques that consider different factors like transaction frequency, available resources, data rate, and intended usage of data.
Devise shared cryptography for communication.
Hybrid encryption techniques.
Standardized cryptographic techniques.
Multi-factor cryptographic solutions.
Traffic filtering techniques.

2.5.4 Cybersecurity Vulnerabilities, Threats, and Attacks in Application Layer

The application layer is the most visible layer of the IoE architecture. It includes all the applications associated with the IoE. Applications are the software programs that allow users to interact with things and access data. These applications can run on various devices including smartphones, tablets, and computers and must be designed for usability and performance. In the application layer, end users have the ability to utilize information from smart devices, enabling personalized services tailored to their specific requirements. The goal of establishing the IoE is to leverage applications that enhance lifestyles and alleviate workloads. An application layer protocol is employed across multiple end systems, facilitating the exchange of packets between programs running on different systems.

The utilization of the DTLS protocol as a secure communication means authorized by CoAP is prevalent. TLS/SSL protocols manage the security of MQTT and

AMQP protocols. Currently, a universal standard for the IoT/IoE application layer [12] is absent, which results in variable security solutions for different application environments. For instance, the 6LoWPAN architecture is utilized for security solutions within some industries. DTLS-based application security architectures underpin the majority of application security architectures that present their security model with CoAP, while certain application security architectures rely on HTTP payload encryption. Furthermore, data sharing is a characteristic of the application layer, and it raises concerns about information disclosure, privacy, and access control, given that each application has multiple users. Therefore, specific authentication mechanisms should be employed to prevent unauthorized access by users for each program. Additionally, mechanisms for data processing and its algorithms are not invulnerable, which could lead to data or information loss and catastrophic damage. Therefore, two factors are taken into account when addressing the security issue in the application layer: authentication condition and key agreement in the heterogeneous network and user privacy protection. Moreover, information security training and effective management practices, specifically regarding password management, play a vital role in maintaining security. The following are some common application layer attacks:

- *Firmware replacement attack:* During an object's operational or maintenance phase, it is common to upgrade its operating system, software, and firmware to leverage new functionalities. However, an attacker can exploit this upgrade process to compromise the object's operational behavior by replacing malicious components as follows:

  SQL injection.
  XSS injection (cross-site scripting attacks).
  Enumeration (CWE/SANS).
  Common vulnerability.
  Phishing attack.
  Sniffing attack.
  Buffer overflow.

2.5.5 Cybersecurity Considerations and Solutions in Application Layer

Although not publicly disclosed, the first attack on the IoE was detected in 2013, targeting an interconnected heating, ventilation, and air conditioning (HVAC) system within an enterprise infrastructure [23]. This allowed the attackers to reach out to a third party that had a keen interest in observing the HVAC operations within the organization. The attackers managed to acquire a user's credentials information and subsequently infiltrated the company's point of sale system with success. There is a need to design a security framework to ensure the protection of both data and devices within the IoE paradigm. A blockchain is one of the decentralized solutions that incorporate hardware security components to address the difficulties related to scalability, security, and latency. The security framework should be decentralized management through the utilization of blockchain and smart contracts for more security. In terms of data transmission security, the processing layer commonly employs homomorphic encryption. However, one challenge associated with this approach is the increased consumption of data. Service-level agreement (SLA) is an effective approach for ensuring the security of processes within the processing layer of the IoE paradigm. Within the processing layer, protocols such as fragmentation redundancy are employed to reduce process vulnerabilities. These protocols achieve this by dividing and assigning data hierarchically to specific target sources, such as cloud nodes or end devices. Moreover, end-to-end process protection frameworks are implemented to facilitate secure data transfer between different layers and devices.

To ensure the security of applications, it is important to take into account the following common best practices:

Develop applications (including web, mobile, and cloud applications) with robust and standardized secure coding practices to minimize the risk of potential attacks.
Check accuracy for input data.
Conduct comprehensive testing of applications (including dynamic, static, and dual testing) to identify vulnerabilities, and promptly address any issues to prevent potential damages and prevent information disclosure.
Employ code signing, also known as coded signatures, to provide customers with assurance regarding the accuracy of the software.
Implement continuous monitoring of important files to prevent unauthorized changes.
Verify the identity and credentials of users through authentication mechanisms focused on ensuring the data security throughout data transfer [24].
Include a certificate signature, digital signature, and certificate chain of a software update package [25].
Deactivate software ports that are not essential for the regular operation.
Utilize a unique encryption key, distinct from other software keys, to validate the integrity of the final software.
Implement a segregation strategy for sensitive software components, such as cryptographic processes, by isolating them from other software components or rating them more.

2.5.6 IoE Business Layer Security

The business layer of IoE is responsible for ensuring compliance with regulations to create a secure and safe environment. Clear policies and protocols are needed to define IoE ecosystem. Cisco's Smart City Framework [26] provides guidelines for secure and responsible IoE-powered smart cities, promoting collaboration and innovation while ensuring security, privacy, and reliability. Adopting such frameworks can establish common standards and best practices for IoE-powered solutions, building trust and paving the way for widespread adoption.

2.5.7 IoE End-to-End Security

Because technology is evolving so rapidly on multiple levels, there is occasionally a lack of in-depth investigation from a security and privacy perspective. This creates gaps and vulnerabilities that can be exploited by attackers. End-to-end security and data protection are critical components of IoE architecture and must therefore be carefully managed, especially when data is transferred across multiple connected devices, applications, and processes. The security framework must be designed to protect the network, data, and applications from unauthorized access, malware, and other threats. Taking into account the costs associated with implementing data security and encryption, these techniques vary in their overhead requirements, ranging from "light" to "medium" to "heavy" depending on the availability of computing power.

2.6 Conclusion

The age of the IoT is being replaced by the age of the IoE, in which everything can communicate with everything. Since the IoE is the evolution of the IoT, it brings with it security challenges (vulnerabilities, threats, and attacks) in addition to those of the IoT. The advent of IoE has undoubtedly expanded the attack surface and the range of activities in our daily lives that are affected. IoE entities suffer from various types of threats, and it is increasing day by day. To protect the security of devices and the privacy of consumers, it is essential to prevent attackers from entering the devices or the network. In this chapter, we have provided a comprehensive overview of cybersecurity issues, threats, and countermeasures in IoE. First, we introduced the IoE entities and the relationships between their four pillar components, as well as the IoE cybersecurity architecture as an evolution of IoT. Then, we defined cybersecurity vulnerabilities, threats, and attacks in different layers of IoE; finally, we defined different cybersecurity considerations and solutions presented in different papers. In this chapter, we have presented a comprehensive survey of IoE and a survey of the potential cybersecurity threats and attacks in IoE and finally proposed an IoE 3D cybersecurity model, which is end-to-end security based on three dimensions of the multi-layered architecture of IoE, including IoE components and entities (first dimension), IoE and its multi-layered ICT architecture (second dimension), and IoE and its enabling technologies (third dimension), in the context of digital transformation. This overview provides important insights for the implementation of future IoE systems. The future research endeavors should prioritize crucial aspects and ensuring effective control capabilities within the realm of the IoE.

References

1. Qureshi KN (2018) New Trends in Internet of Things, Applications, Challenges, and Solutions. TELKOMNIKA (Telecommunication Computing Electronics and Control) 16 (3):1114-1119. doi:https://doi.org/10.12928/TELKOMNIKA.v16i3.8483
2. Khan WZ, Rafique W, Haider N, Hakak S, Imran M (2022) Internet of Everything: Enabling Technologies, Applications, Security and Challenges. TechRxiv Preprint. doi:https://doi.org/10.36227/techrxiv.21341796.v1
3. Mohanty SP, Yanambaka VP, Kougianos E, Puthal D (2020) PUFchain: A hardware-assisted blockchain for sustainable simultaneous device and data security in the internet of everything (IoE). IEEE Consumer Electronics Magazine 9 (2):8-16. doi:https://doi.org/10.1109/MCE.2019.2953758
4. Ryoo J, Kim S, Cho J, Kim H, Tjoa S, DeRobertis C IoE security threats and you. In: 2017 International Conference on Software Security and Assurance (ICSSA), 2017. IEEE, pp 13–19. doi: 10.1109/ICSSA.2017.28
5. Konev A, Shelupanov A, Kataev M, Ageeva V, Nabieva A (2022) A survey on threat-modeling techniques: protected objects and classification of threats. Symmetry 14 (3):549. doi:https://doi.org/10.3390/sym14030549
6. Li L, Pahlevanzadeh B (2021) Evaluation of the trust values among human resources in the enterprise cloud using an optimization algorithm and fuzzy logic. Kybernetes 51 (6):2008-2029. doi:https://doi.org/10.1108/K-04-2021-0280
7. Swathi GC, Kumar GK, Kumar AS (2022) Estimating Botnet Impact on IoT/IoE networks using Traffic flow Features. Computers and Electrical Engineering 102:108209
8. Qureshi KN, Alhudhaif A, Haider SW, Majeed S, Jeon G (2022) Secure Data Communication for Wireless Mobile Nodes in Intelligent Transportation Systems. Microprocessors and Microsystems:104501. doi:https://doi.org/10.1016/j.micpro.2022.104501
9. HaddadPajouh H, Dehghantanha A, Parizi RM, Aledhari M, Karimipour H (2021) A survey on internet of things security: Requirements, challenges, and solutions. Internet of Things 14:100129. doi:https://doi.org/10.1016/j.iot.2019.100129
10. Pahlevanzadeh B, Koleini S, Fadilah SI Security in IOT: Threats and vulnerabilities, layered architecture, encryption mechanisms, challenges and solutions. In: Advances in Cyber Security: Second International Conference, ACeS 2020, Penang, Malaysia, December 8-9, 2020, Revised Selected Papers, 2021. Springer, pp 267–283. doi:10.3390/electronics11203330
11. Chen K, Zhang S, Li Z, Zhang Y, Deng Q, Ray S, Jin Y (2018) Internet-of-Things security and vulnerabilities: Taxonomy, challenges, and practice. Journal of Hardware and Systems Security 2:97-110
12. Zhen-hua D, Jin-tao L, Bo F A taxonomy model of RFID security threats. In: 2008 11th IEEE International Conference on Communication Technology, 2008. https://doi.org/10.1109/ICCT.2008.4716242
13. Stan O, Bitton R, Ezrets M, Dadon M, Inokuchi M, Ohta Y, Yagyu T, Elovici Y, Shabtai A (2020) Extending attack graphs to represent cyber-attacks in communication protocols and modern it networks. IEEE Transactions on Dependable and Secure Computing 19 (3):1936-1954. doi:https://doi.org/10.1109/TDSC.2020.3041999
14. Sadhu PK, Yanambaka VP, Abdelgawad A (2022) Internet of Things: Security and Solutions Survey. Sensors 22 (19):7433. doi:https://doi.org/10.3390/s22197433
15. Atamli AW, Martin A Threat-based security analysis for the internet of things. In: 2014 International Workshop on Secure Internet of Things, 2014. IEEE, pp 35–43. doi:https://doi.org/10.1109/SIoT.2014.10
16. Lu Y, Da Xu L (2018) Internet of Things (IoT) cybersecurity research: A review of current research topics. IEEE Internet of Things Journal 6 (2):2103-2115. doi:https://doi.org/10.1109/JIOT.2018.2869847

17. Shamsoshoara A, Korenda A, Afghah F, Zeadally S (2020) A survey on physical unclonable function (PUF)-based security solutions for Internet of Things. Computer Networks 183:107593. doi:https://doi.org/10.1016/j.comnet.2020.107593
18. Nawir M, Amir A, Yaakob N, Lynn OB Internet of Things (IoT): Taxonomy of security attacks. In: 2016 3rd international conference on electronic design (ICED), 2016. IEEE, pp 321–326. doi:https://doi.org/10.1109/ICED.2016.7804660
19. Rehman M, Javed IT, Qureshi KN, Margaria T, Jeon G (2022) A Cyber Secure Medical Management System by Using Blockchain. IEEE Transactions on Computational Social Systems:1–14. doi:https://doi.org/10.1109/TCSS.2022.3215455
20. Un Nisa K, Alhudhaif A, Qureshi KN, Hadi HJ, Jeon G (2022) Security Provision for Protecting Intelligent Sensors and Zero Touch Devices by using Blockchain Method for the Smart Cities. Microprocessors and Microsystems:104503. doi:https://doi.org/10.1016/j.micpro.2022.104503
21. Borgohain T, Kumar U, Sanyal S (2015) Survey of security and privacy issues of internet of things. arXiv preprint arXiv:150102211. doi:https://doi.org/10.48550/arXiv.1501.02211
22. Behera TM, Mohapatra SK, Samal UC, Khan MS, Daneshmand M, Gandomi AH (2019) Residual energy-based cluster-head selection in WSNs for IoT application. IEEE Internet of Things Journal 6 (3):5132-5139. doi:https://doi.org/10.1109/JIOT.2019.2897119
23. Tariq U, Ahmed I, Bashir AK, Shaukat K (2023) A Critical Cybersecurity Analysis and Future Research Directions for the Internet of Things: A Comprehensive Review. Sensors 23 (8):4117. doi: https://doi.org/10.3390/s23084117
24. Raza S, Helgason T, Papadimitratos P, Voigt T (2017) SecureSense: End-to-end secure communication architecture for the cloud-connected Internet of Things. Future Generation Computer Systems 77:40-51. doi:https://doi.org/10.1016/j.future.2017.06.008
25. Yang G, Xu J, Chen W, Qi Z-H, Wang H-Y (2010) Security characteristic and technology in the internet of things. Nanjing Youdian Daxue Xuebao(Ziran Kexue Ban)/ Journal of Nanjing University of Posts and Telecommunications (Natural Nanjing University of Posts and Telecommunications) Natural 30 (4)
26. Falconer, Gordon, Mitchell S (2012) Smart city framework. Cisco Internet Business Solutions Group (IBSG) 12 no. 9 (2012): 2-10

Chapter 3
Attack Detection Mechanisms for Internet of Everything (IoE) Networks

Raja Waseem Anwar and Kashif Naseer Qureshi

3.1 Introduction

An innovative new development that is changing people's lives is the Internet of Everything (IoE), which is a superset of the Internet of Things (IoT), people, data, processes, and things. The devices are connected by using advanced technologies and increase the relevance, value, and capacity of the network for data sharing and communication services. The IoE encapsulates the concept of that how a human connects with objects and the world where the cyber and physical systems intersect and work together. Additionally, these connected ecosystems rely on embedded electronic devices like sensors and actuators for data collection and information dissemination. The goal of connecting people and devices worldwide is to provide cost-effective, fast data communication services. Every decade since its inception has contributed in a different way to the development of the Internet. In recent time, the penetration rate is about 40.4% of the world's population [1], whereas the concept of "more data, more collaboration, and more complex systems of interactions" has gained importance. It alludes to the notion that, in order to facilitate communication, everything must be equipped with sensors and transmitters [2]. A network of online-connected devices that enabled cross-cultural communication, learning, and cognition would eventually exist. It is said that the IoE is a movement that will

R. W. Anwar (✉)
Department of Computer Science, German University of Technology in Oman, Muscat, Sultanate of Oman
e-mail: raja.anwar@gutech.edu.om

K. N. Qureshi
Department of Electronic & Computer Engineering, University of Limerick (UL), Limerick, Ireland
e-mail: kashifnaseer.qureshi@ul.ie

K. Naseer Qureshi et al. (eds.), *Cybersecurity Vigilance and Security Engineering of Internet of Everything*, Internet of Things,
https://doi.org/10.1007/978-3-031-45162-1_3

eventually fundamentally alter how we live. The IoE is a conceptual framework that builds on the IoT and Machine-to-Machine (M2M) communication to depict a more complex system that also includes processes, data, and things [3].

It is possible to gain network links that are more useful and important than ever before. To achieve this, each piece of information can be transformed into a series of steps that result in the development of new functions and processes. IoE enables people, businesses, and governments to generate new economic opportunities for themselves. The IoE includes sophisticated information-sharing methods, but in order for it to be broadly adopted across all industries, it must be developed and deployed with the appropriate safety precautions in place from the outset. IoE uses sensors to connect a variety of devices to the Internet so that data may be gathered and analyzed intelligently. Technologies that are intelligent, interactive, and instructional are rapidly altering the modern world. Examples of these technologies include smartphones, expert systems, and data processing systems [4]. The IoE has been seen as facilitating integration and communication among intelligent items as well as advancing new technologies and uses. The term "intelligent object" refers to a network of multiple objects that have orchestration and convergence qualities as well as visibility across earlier separate systems [5].

As a result of cyberattacks, emerging threats, and recently found weaknesses, the IoE networks now face new issues. In recent years, research initiatives have embraced a variety of strategies, including cyber-smart, cybersecurity, and cyber-safety. The importance of economic and commercial environment plays a crucial role in the execution of security. According to Cisco statistics, the frequency of attacks between 100 Gbps and 400 Gbps grew by 76% from 2018 to 2019, and the number of distributed denial-of-service (DDoS) attacks globally is anticipated to double from 7.9 million in 2018 to 15.4 million in 2023 [6].

This chapter examines the security threats, challenges, risks, and mitigation strategies specifically designed for the IoE landscape. New security threats and cyberattacks are a significant concern for the IoE networks due to heterogeneous systems and open systems. As a result, the systems are compromised and vulnerable, and user privacy is compromised. It is essential to protect the IoE devices and establish strong security frameworks and safeguard the networks against these dangers and attacks.

3.2 Definition, Elements, and Applications of IoE Networks

The IoE network is a cutting-edge idea that builds a system that benefits several parties, including enterprises, governments, and individuals, and is more effective, intelligent, and integrated. IoT networks only focus on tangible objects, whereas the IoE seeks to connect everything to the Internet in order to assist intelligent service delivery and decision-making [7]. Increased context awareness, energy independence, and processing power are highly valued by the IoE. With the help of the IoE networks, people, processes, and objects can interact in new ways, producing

Fig. 3.1 Main components of IoE

valuable data that can be used to improve decision-making. The primary components of IoE are depicted in Fig. 3.1. It consists of numerous interconnected systems, applications, and objects. IoE might facilitate the utilization of all crucial data and processes, thereby increasing the value and relevance of IoT for people. The major objectives of IoE are to increase operational effectiveness, provide new economic prospects, and improve people's lives [8].

3.2.1 IoE Network Components

IoE networks are made up of four essential parts that cooperate to allow for effective communication and data sharing [7]:

- *People*: This component describes the users who communicate across the network, such as people, businesses, and governments. Using applications and interfaces, users can access information and manage a variety of networked devices.
- *Processes*: Within the IoE network, processes govern how data is gathered, examined, and used. To increase general effectiveness and productivity, this involves data management, decision-making algorithms, and the automation of various operations.
- *Data*: IoE networks are built on data since it powers both applications and decision-making. It includes both unprocessed sensor data and processed and analyzed data that is used to inform decisions and uncover new information. IoE

networks receive data from a variety of sources, including devices, people, and other systems. This data might be structured or unstructured.

- *Things*: The physical items or gadgets linked to the IoE network are referred to as "things." These gadgets may consist of sensors, actuators, cameras, and other data-gathering and data-transmitting smart items. Additionally, they can interface with other network users or devices and carry out a variety of tasks.

Integration of intelligent networks and services enables the realization of each of these possibilities. Connection, network economy, collaborative experience, and the IoE are the four main phases of the development of the Internet. Each of these phases has been defined by the enormous growth of larger businesses and, more generally, of establishments. Services that are available only during the connectivity phase include email, web surfing, content searching, and other comparable activities [9]. The world has changed significantly since the year 2000 because social media services, video, mobile, and cloud computing have made it simpler for individuals to collaborate. This era of the Internet's development, known as the IoE, is concerned with connecting people, processes, data, and objects.

3.2.2 Applications of IoE Networks

IoE networks have several uses in a variety of industries and provide several advantages like improved productivity, cost savings, and user experiences. Several well-known uses are the following [10]:

- *Smart Cities*: IoE networks contribute to the development of smart cities by regulating traffic, reducing energy consumption, providing real-time data on public transit, and enhancing public safety through surveillance and emergency response systems.
- *Healthcare*: In the field of medicine, IoE networks allow for telemedicine, tracking of medical equipment, and remote patient monitoring. Additionally, they help healthcare professionals share data, which enhances patient care and advances medical investigation.
- *Agriculture*: By delivering real-time data on soil, weather, and crop conditions, IoE networks can revolutionize farming and enable precision agriculture as well as the efficient use of resources like water and fertilizers.
- *Manufacturing*: IoE networks enable automation, preventive maintenance, and real-time monitoring of production processes in the industrial sector, facilitating the implementation of Industry 4.0. As a result, productivity rises, downtime decreases, and product quality improves.
- *Energy*: By allowing smart grids, which can automatically balance supply and demand, include renewable energy sources, and keep track of the entire infrastructure, IoE networks help optimize energy use and distribution.

- *Transportation*: The development of driverless vehicles, intelligent traffic management, and vehicle-to-vehicle and vehicle-to-infrastructure (V2I) communication is a possible use of IoE networks in the transportation sector.
- *Retail*: IoE networks can enhance customer experiences in the retail sector by providing customized marketing, real-time inventory management, and intelligent payment systems.
- *Environmental Monitoring*: IoE networks can support the monitoring and analysis of environmental parameters like water pollution, air quality, and natural disasters, enabling better decision-making and prompt interventions.

By fostering a more connected, effective, and intelligent environment, the IoE networks have the potential to revolutionize a number of industries and enhance the quality of life. To fully take advantage, it is necessary to address the security issues brought on by the extensive use of IoE networks. IoE appears to be a step-up from the IoT, which is concerned with connecting physical objects and utilizing existing communication technology in order to increase utility value. By tying up actual items and leveraging one of the most recent communication technologies, the IoE seeks to achieve this. We commonly discuss the IoE while increasing the capabilities of the IoT [11].

Some of these include increased processing power, increased environmental awareness, an independent energy supply and increased recruitment and use of new connected information types. Because it covers both vertical and horizontal products and services, such as wired and wireless networks, the IoE is unique in its heterogeneity which is also crowded with a wide range of items, from simple toys to complex computing apparatus.

The IoE links people, data, objects, and processes to make connectivity simpler and more pervasive than ever before as the number of IoE devices increases and begins to play a significant part in people's daily lives. While data is used to inform decisions, process provides the appropriate to the right person. By enhancing the intelligence and automation of commercial and industrial operations, IoE-enabled gadgets can enhance people's lives. Although technological advancements are the primary driver of economic progress, they also increase cyberattacks on the IoE across a range of sectors and companies. Cyberattacks are a key factor to take into account when making investments in cybersecurity since they encompass all risks to information systems in the wireless IoE. IoE may have many advantages, but it also has significant security threats [12].

3.3 Challenges and Issues

With a number of benefits, the IoE network brings new challenges such as scalability, infrastructure, and security concerns. Although advancements in technology are the primary factor influencing economic growth, they also increased the frequency of cyberattacks and the number of user privacy violations. Cyberattacks are the main challenge and cause of the degradation of network services. Security is also an important factor to take into account because they encompass all dangers to information systems in the wireless IoE.

3.4 IoE Security Requirements

The most important issue for the IoE is security and needs attention. Any security breach can have disastrous consequences, including loss of money, information, and personal safety due to the insertion of inaccurate data into the system, which disrupts various activities and decision-making. IoE is more important than IoT, in terms of security, network congestion, privacy, and energy consumption due to the direct involvement of user's social behavior. The IoT is a network of interconnected things with the ability to sense their surroundings and respond appropriately. It is possible to grow an intelligent and safe IoE system without compromising the services and communication processes [13]. It is possible to build up this safe IoE infrastructure without sacrificing security or intelligence. These environments are built up in a way that leaves them subject to a variety of dangers and serious security threats, especially given their shoddy connectivity and open data interchange.

The primary objective of IoE security is to protect networks, data, and physical assets from known and unknown vulnerabilities, threats, and attacks. A significant amount of information is produced by a diverse range of devices and utilized to inform decisions. Additionally, the acquired data is regarded as the most valuable asset and requires adequate protection to preserve its availability and authenticity. Integrity is the belief in the veracity of a system's resources, which ensures that actions are taken by those who are authorized to and are meant to be taking them. Table 3.1 lists the numerous security requirements that must be taken into account by the different IoE components during the design and authentication phases [14–16].

The IoE must safeguard its data's integrity and take the appropriate safeguards to stop enemies from assaulting and listening in on communications. It is also crucial to safeguard the confidentiality of data and system communications and ensure uncompromised security in order to provide data and transactions a sense of availability, authenticity, and validity. The IoE has many benefits and a bright future, but it also poses some serious challenges and issues that must be addressed. These issues and threats are still present.

3.5 Security Attacks in IoE

The IoE networks have a significant impact on our daily activities as there are more and more IoE devices all around us, since attacks on IoE devices can directly affect the privacy of end users. The link that IoE creates between the real world and

Table 3.1 IoE: security requirements

IoE networks' security requirements	Description
Confidentiality	Efforts are made to prevent unauthorized access, and as a result, the data is safe and only accessible to authorized users
Integrity	End-to-end encryption and digital signatures can be used in an IoE setting to ensure data integrity
Availability	Enables rapid and trustworthy usage of data, tools, and services
Authentication	Every object in the IoE must be able to identify and confirm other objects in order to function the network of interconnected things, people, services, providers, and processing units
Authorization	Only permission is allowed to users to access the provided tools and services
Non-repudiation	A cybersecurity condition that provides evidence of what entities have done in IoE networks is non-repudiation
Data freshness	Data reliability enables the assurance that all data generated by devices is up to date, time-stamped, and free from tampering by an adversary who might have modified the data or retransmitted prior communications
Anonymity	Anonymity ensures the data is safe and inaccessible to possible enemies
Scalability	Is the system's capacity to maintain its current devices and service set while adding new ones
Attack resistance	Attack deterrence provides resistance to several potential assaults

cyberspace raises the risk of cyberattacks that target IoE devices. It is more challenging than ever for businesses to stay on top of the most recent security concerns since the IoT is evolving so swiftly. Manufacturers of connected products frequently overlook security issues in the architecture or design of the system because they place an undue premium on utility and remote control. One of the current concerns is the possibility of state-sponsored cyberwar, in addition to viruses, worms, malware, spyware, botnets, spam, DDoS, ransomware, advanced persistent threats (APT), identity theft, phishing, and hacktivism. Attackers are amplified to use wireless network flaws to their advantage. Additionally, almost all operators continue to face the serious possibility of critical infrastructure failure. The devastating attacks might take place if the nodes are installed maliciously [17, 18].

- *The Sybil attack*: An attacker may try to adopt a different identity close to another node after capturing control of one. A rogue node poses multiple other nodes during this type of attack [16].
- *Node replication attack*: A node replication attack may be used to retrieve the credentials for this node from the memory of a smart device that has been compromised by the adversary. This makes it possible to make a replica of this node and position it "near another node" in the network.
- *Sinkhole attack*: A node can use the sinkhole attack to broadcast to the nodes around it the shortest number of hops possible and the best available path to the target node. Network traffic can reach the sinkhole node, which serves as the destination node, by convincing the nearby nodes to use these paths.

- *Wormhole attack*: By tunneling the transmitted packets or information between the two remote nodes, the adversary can cause one remote node to discover a new neighbor node that is actually a mirror image of the other. This attack can be carried out in a variety of methods, such as changing, sniffing, and dropping.

In addition to the threats already mentioned, there are a number of other attacks that must be protected against in an IoE setting. Beyond security and privacy issues, IoE devices are susceptible to new kinds of attacks because of how they function. A brief summary of a few of these attacks is given below [19, 20].

- *Replay attack*: By using the same information that was provided during the interaction, an adversary attempts to deceive another legal body in this kind of attack. It's referred to as a "replay attack."
- *Man-in-the-middle attack*: Using this method, an attacker is able to intercept messages while they are being sent. The substance of the communications can then be modified or deleted, or even malicious material can be rapidly added. By doing this, the recipient is kept in the dark about these issues and will therefore treat any messages it receives as though they were received with permission.
- *Impersonation attack*: To fool other entities operating in an IoE environment, the attacker may attempt to create phony communications that appear to have come from a source entity as part of this attack.
- *Insider attack*: An authorized insider user within the company (or, in the event of access control, a trusted authority) may initiate this kind of attack. Once registered users, smart devices, and fog servers have been enrolled or registered, the attacker has access to the sensitive data and can then launch additional attacks, such as impersonation attacks.
- *Forward and backward security*: An access control approach must prevent any new communications from being collected when a smart Internet of Things device or user exits an Internet of Everything environment. In a similar way, a new Internet of Things smart device or user can join the ecosystem, but they must be prevented from accessing any messages that have already been sent.
- *On- or offline attacks (guessing)*: An adversary may be able to successfully guess the credentials (password and biometrics) of a registered user in an access control scheme utilizing intercepted messages and the stored credentials in the system using a stolen-verifier attack using the user's mobile device, whether the user is logged in online or offline. This could happen both online and offline for the user.
- *Ephemeral secret leakage (ESL) attack*: According to the current de facto CK-adversary threat model, an attacker shouldn't be able to determine the session key that is established between two interacting entities during the access control process, even if they are successful in learning the short-term secrets through session hijacking attempts. The opponent won't be able to accomplish that until the long-term secrets have also been made public [21].

- *Malicious code injection attack*: When a malicious code is injected into node's memory with this type of attack and then executed, the attacker has complete control over the network.
- *Jamming attack*: Attacker uses an air interface and the tag reader which disrupts long-distance transmitter communication.
- *DoS attack*: This attack disrupts the target network, prohibiting authorized entities from using it.
- *Malicious script attack*: The user needs access to malicious scripts or programs in order to carry out this kind of attack.
- *Malicious traffic classification*: Prior to offering protection against cyberattacks for both individuals and companies, network traffic must first be examined and categorized in order to detect anomalies and malicious attacks. Because classifying harmful traffic is such a crucial activity, a sizable number of researchers have concentrated on creating more effective classification algorithms by utilizing AI.
- *Unsecured open wireless communication for remote access*: This might lead to an interception, in which case an attacker would take advantage of it to have remote access to a captured device to launch them cyberattack.

If the area where the smart devices are located is unattended, an attacker who has the ability to physically grab one or more of them might be able to do so. It is likely that an attacker compromises the security of communication between the non-compromised nodes in the IoE environment if they are able to retrieve the credentials that are stored on the compromised devices. Fig. 3.2 provides a summary of these attacks.

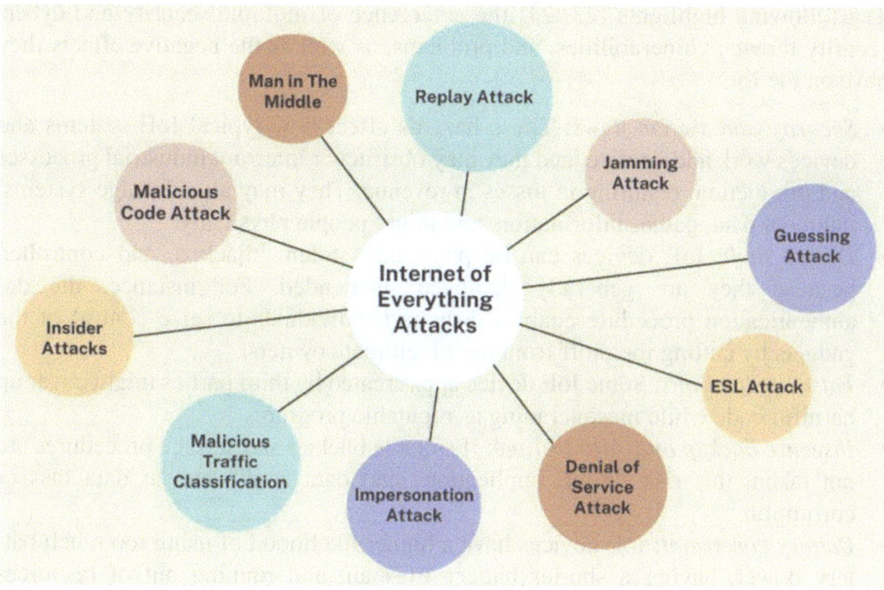

Fig. 3.2 IoE environment (potential attacks)

3.6 IoE: Security Vulnerabilities

IoE systems may be vulnerable in several ways, which could impair their ability to connect, work, and deliver accurate information. Security risks and weaknesses are increasing. These vulnerabilities include the following [22]:

- *Network vulnerability*: IoE systems lack or have not adopted basic security measures, making them vulnerable to multiple wired and wireless communication and connection attacks.
- *Platform vulnerability*: IoE devices are not kept secure and up to date due to a lack of regular software and firmware updates and security patches.
- *Application vulnerability*: Applications that haven't been tested and examined for coding or compatibility issues could affect how well IoE systems and devices work.
- *Security vulnerability*: If new security measures are installed without being thoroughly evaluated, IoE devices and system might not perform as well.
- *Update vulnerability*: IoE is prone to update problems that could result in deleting data that hasn't been saved, suspending a process that is already running, etc.
- *Management vulnerability*: Resulting from a lack of security rules, procedures, policies, and informed planning.

3.7 Security Risks in IoE

The following highlights [23, 24] the emergence of multiple security and cybersecurity threats, vulnerabilities, and problems, as well as the negative effects they have on the IoE:

- *Security and system flaws*: These hazards affect how typical IoE systems and devices work and operate, and they may obstruct or interrupt industrial processes and production, resulting in losses in revenue. They may also damage systems, intercept data, gather information, and injure people physically.
- *Device theft*: IoE devices can be physically stolen, hijacked, and controlled because they are generally deployed unattended. For instance, the de-authentication procedure enables dishonest individuals to seize control of the gadgets by cutting them off from their legitimate owners.
- *False applications*: Some IoE device apps created by third parties might cover up harmful code while masquerading as reputable programs.
- *Insecure backup and data storage*: If suitable backup and storage procedures are not taken, this risk to IoE applications and data could lead to data loss or corruption.
- *Battery constraint*: IoE devices have a higher likelihood of using too much battery power, having a shorter battery lifespan, and running out of resources because they are resource-constrained by nature.

- *Non-backed communication:* If the deployed apps or devices are mission-critical in nature, communications being intercepted or IoE devices losing contact could have a severe effect on the decision-making process.
- *People's security*: Because people are typically easy victims, they are the weakest link in the security chain. Just as much as the basic network architecture needs to be hardened and security algorithms, tools, and processes offered, the system's users also need to be secured.
- *Things security*: Despite the fact that the data in the IoE environment is encrypted, a small number of IoE devices might be exploited to construct several identities, and device heterogeneity could provide the attacker with crucial information.
- *Process security*: Due to the recognized weaknesses in the various data processing protocols, it is equally critical and challenging to secure processes in the IoE [25].
- *Data security*: Data security is one of the biggest concerns in the highly networked IoE paradigm, where everything gathers and exchanges data online.

3.8 Privacy Challenges in IoE

The IoE, which automates and enhances industrial processes, improves people's lives. IoE is a dynamic ecosystem made up of numerous linked devices and services that exchange data and information. Different factors in the IoE can lead to security and privacy issues. IoT is a key component of IoE; therefore, there are several levels with various attack types that can be dealt with in various ways.

Security is crucial with the existing decentralized infrastructure; nevertheless, deployed devices are more vulnerable to attacks because of their low energy and computing capabilities. Additionally, IoE systems are frequently deployed in remote locations, making them vulnerable to physical assaults. IoE devices collect, process, and transmit sensitive data via a network. The user should be aware of the private data being processed, and attackers should have adequate protection for this data. Each domain in the IoE ecosystem has its own trust, security, and privacy challenges, and the environment is thought to be vibrant. Several privacy concerns include the following [26]:

- *User confidentiality and data protection*: The IoE connects objects to enable data sharing while preserving user privacy and security.
- *User authentication and identity management:* For identity and authentication, the IoE uses a range of methods and tools. The environment is uniquely identified using identity management techniques, and the identity establishment between objects in the IoE environment is guaranteed and verified using authentication procedures.
- *Trust management and policy integration:* Establishing trust among communication entities is necessary because of the ambiguous IoE environment. User trust and object-to-object trust are the two perspectives on trust in the IoE.

- *Authorization and access control:* After something or someone has been recognized, it is possible to ascertain whether they have permission to utilize a resource by using authorization. There are numerous variables that influence whether someone has access to resources or not.
- *End-to-end security:* It is also very important to secure the locations where IoE devices connect to Internet hosts. To provide complete end-to-end IoE security, packet codes for authentication and encryption are insufficient. Session keys and algorithms need to be used securely if end-to-end security is to be achieved.
- *Attack-resistant security solution*: IoE devices come in a wide variety, have varied memory capacities, and have constrained computing resources. As a result, attacks ought to be discouraged, and proper security countermeasures ought to be put in place.

It is equally crucial to protect users' privacy in the Internet of Everything because if an IoE device is compromised, adversaries can exploit it in a variety of ways to carry out attacks and other illegal acts. For example, a compromised IoE device may endanger user privacy by disclosing sensitive data. Additionally, the adversary will be able to disable the device or even threaten the victim using ransomware [27].

3.9 Attack Detection and Countermeasures

Because of how IoE applications and networks are utilized, security needs are among the most crucial non-tangible criteria that should be taken into consideration at an early stage. This is due to IoE's versatility. With the help of the new IoE technology, sensors and control systems with physical network connectivity and processing capabilities may now produce, exchange, and consume data with little assistance from humans. IoE security, nevertheless, differs from typical security in that it necessitates novel and imaginative approaches to safeguard gadgets and programs while taking into account factors like constrained resources, a distributed architecture, and a variety of locations. The IoE faces specific challenges, including unreliable communication, a lack of data, and insufficient privilege protection.

Effective security measures must be put in place and kept up to date in order to protect the IoE. In order to prevent malicious or unauthorized access, a robust multifactor authentication mechanism must be in place in addition to the procedures for identifying and confirming individuals. In actuality, protecting IoE systems is a difficult undertaking. It is not, however, an insurmountable feat. In order to secure the IoE and its components, various cryptographic, non-cryptographic, and Artificial Intelligence (AI)-based solutions were offered. The following are some of the assault detection strategies for IoE that are emphasized [28, 29]:

- *Cyber threat intelligence*: The Internet of Everything (IoE) risks and threat actors are the foundation of Cyber Threat Intelligence (CTI). Based on the concept of an Advanced Persistent Threat (APT), this information would aid in the early detection and prevention of dangerous cyber-events.

- *Active response (detection and prevention)*: Active response necessitates the adoption of easier-to-setup and operates detective and preventative security techniques and platforms. These programs and platforms offer additional security defenses.
- *Artificial Intelligence (AI)-based detection*: Using machine learning (ML)-based techniques to implement AI-based attack detection systems for IoE ensures high precision [30].
- *Cryptographic solutions and protocols*: The most common use of cryptographic protocols is to verify the identification of a person or device. These protocols rely on cryptographic techniques at their core, and they are used for authentication. However, these methods are not economically viable due to the resource limitations of IoE devices like sensors and actuators.
- *Intrusion detection systems (IDS) and firewalls*: Implementing intrusion detection and prevention systems that boost security in IoE environments is crucial. These systems should use a signature- and anomaly-based detection.
- *Honeypot security solutions*: In order to build robust and sophisticated security mechanisms, honeypots are utilized in conjunction with IDS and firewalls for the quick discovery and mitigation of threats.
- *Online and offline security*: This type of security relies on a cryptographic process in which the message is first offline encrypted, and the results are preserved before transmission and identification of the destination, decreasing the delay and preserving the device's computational cost. Using the first phase's generated results that were stored, the second phase will be completed online [31].
- *IoE and blockchain:* Blockchain technology is the most suitable alternative for data security in this context because the different applications that make up the IoE ecosystem are interconnected and generate a large volume of data. The IoE offers an extensive range of potential applications, which has prompted the creation of various unique blockchain variants, each of which is based on a different consensus algorithm.

The IoE opens up opportunities for individuals, communities, and nations that have never been feasible before by recognizing the importance of connecting people, processes, data, and things as a whole. Systems built on the IoE include complex architectures, communication-intensive system layers, and several other vulnerabilities that make them attractive targets for hackers.

3.10 Conclusion

The IoE, which consists of numerous gadgets connected to one another, is gaining popularity. It is difficult to offer security and privacy as a result. Security procedures and policies will be an integral part of IoE's foundation. Identity-based security is a better solution for safeguarding the IoE as compared to perimeter-based security. IoE includes, among other things, the ongoing detection of potential weaknesses, the practice of security education, and the ongoing evaluation of security laws. In this chapter, we examine the main attacks, problems, and weaknesses of the IoE as well as how to defend against them, with a focus on security, privacy, and risk considerations. However, the vast majority of newly discovered security mechanisms and privacy-preserving techniques could not be broadly adopted in various IoE domains due to the heterogeneous environment. These issues won't be resolved either unless there is a set standard that manufacturers must adhere to when developing new technologies or devices with security and privacy in mind. IoE becomes less beneficial as a result of other areas. However, both academics and professionals in the industry can benefit from using this chapter as a resource and a point of reference.

References

1. Buczak AL, Guven E (2015) A survey of data mining and machine learning methods for cyber security intrusion detection. IEEE Communications surveys & tutorials 18 (2):1153–1176
2. Evans D (2012) Internet of everything: Harnessing an exponentially more powerful internet. URL: https://blogs cisco com
3. Langley DJ, van Doorn J, Ng IC, Stieglitz S, Lazovik A, Boonstra A (2021) The Internet of Everything: Smart things and their impact on business models. Journal of Business Research 122:853–863. https://doi.org/10.1016/j.jbusres.2019.12.035
4. Miraz MH, Ali M, Excell PS, Picking R (2015) A review on Internet of Things (IoT), Internet of everything (IoE) and Internet of nano things (IoNT). 2015 Internet Technologies and Applications (ITA): 219–224. https://doi.org/10.1109/ITechA.2015.7317398
5. Anwar RW, Qureshi KN, Nagmeldin W, Abdelmaboud A, Ghafoor KZ, Javed IT, Crespi N (2022) Data Analytics, Self-Organization, and Security Provisioning for Smart Monitoring Systems. Sensors 22 (19):7201
6. Cisco U (2021) Cisco annual internet report (2018–2023) white paper. 2020. Acessado em 10 (01):1–35
7. Bokhari S, Hamrioui S, Aider M (2022) Cybersecurity strategy under uncertainties for an IoE environment. Journal of Network and Computer Applications 205:103426. https://doi.org/10.1016/j.jnca.2022.103426
8. Arroyo-Figueroa G, Rojas-Gonzalez I, Hernández-Aguilar JA (2022) A Compressive Compilation of Cyber Security for Internet of Energy (IoE). In: Research Anthology on Smart Grid and Microgrid Development. IGI Global, pp 883–910. https://doi.org/10.4018/978-1-6684-3666-0.ch039
9. Mithcell S, Villa N, Stewart-Weeks M, Lange A (2013) The Internet of Everything for Cities: Connecting People, Process, Data, and Things To Improve the 'Livability' of Cities and Communities. San Jose: Cisco
10. Ma H, Zhang Y, Shen M (2021) Application and prospect of supercapacitors in Internet of Energy (IOE). Journal of Energy Storage 44:103299. https://doi.org/10.1016/j.est.2021.103299
11. Naseem S, Alhudhaif A, Anwar M, Qureshi KN, Jeon G (2022) Artificial general intelligence-based rational behavior detection using cognitive correlates for tracking online harms. Personal and Ubiquitous Computing. https://doi.org/10.1007/s00779-022-01665-1
12. Alamer M, Almaiah MA Cybersecurity in Smart City: A systematic mapping study. In: 2021 International Conference on Information Technology (ICIT), 2021. IEEE, pp 719–724. https://doi.org/10.1109/ICIT52682.2021.9491123
13. Liu Y, Dai H-N, Wang Q, Shukla MK, Imran M (2020) Unmanned aerial vehicle for internet of everything: Opportunities and challenges. Computer communications 155:66–83. https://doi.org/10.1016/j.comcom.2020.03.017

14. Anwar RW, Ali S (2022) Smart Cities Security Threat Landscape: A Review. Computing and Informatics 41(2):405–423. https://doi.org/10.31577/cai_2022_2_405
15. Bera B, Das AK, Obaidat MS, Vijayakumar P, Hsiao K-F, Park Y (2020) AI-enabled blockchain-based access control for malicious attacks detection and mitigation in IoE. IEEE Consumer Electronics Magazine 10 (5):82–92. https://doi.org/10.1109/MCE.2020.3040541
16. Anwar RW, Abdullah T, Pastore F (2021) Firewall best practices for securing smart healthcare environment: A review. Applied Sciences 11 (19):9183. https://doi.org/10.3390/app11199183
17. Bello AD, Lamba O (2020) How to Detect and Mitigate Sinkhole Attack in Wireless Sensor Network (WSN). Int J Eng Res Technol 9
18. Anwar RW, Bakhtiari M, Zainal A, Qureshi KN (2016) Wireless sensor network performance analysis and effect of blackhole and sinkhole attacks. Jurnal Teknologi 78 (4–3)
19. Anwar RW, Zainal A, Outay F, Yasar A, Iqbal S (2019) BTEM: Belief based trust evaluation mechanism for wireless sensor networks. Future generation computer systems 96:605–616
20. Sah DK, Amgoth T, Cengiz K (2022) Energy efficient medium access control protocol for data collection in wireless sensor network: A Q-learning approach. Sustainable Energy Technologies and Assessments 53:102530. https://doi.org/10.1016/j.seta.2022.102530
21. Abosata N, Al-Rubaye S, Inalhan G, Emmanouilidis C (2021) Internet of things for system integrity: A comprehensive survey on security, attacks and countermeasures for industrial applications. Sensors 21 (11):3654. https://doi.org/10.3390/s21113654
22. Abdalla M, Barbosa M, Bradley T, Jarecki S, Katz J, Xu J Universally composable relaxed password authenticated key exchange. In: Advances in Cryptology–CRYPTO 2020: 40th Annual International Cryptology Conference, CRYPTO 2020, Santa Barbara, CA, USA, August 17–21, 2020, Proceedings, Part I 40, 2020. Springer, pp 278–307. https://doi.org/10.1007/978-3-030-56784-2_10
23. Laitinen A, Niemelä M, Pirhonen J (2019) Demands of dignity in robotic care: Recognizing vulnerability, agency, and subjectivity in robot-based, robot-assisted, and teleoperated elderly care. Techné: Research in Philosophy and Technology 23(3):366–401. https://doi.org/10.5840/techne20191127108
24. Frikha T, Chaari A, Chaabane F, Cheikhrouhou O, Zaguia A (2021) Healthcare and fitness data management using the IoT-based blockchain platform. Journal of Healthcare Engineering 2021. https://doi.org/10.1155/2021/9978863
25. Yaacoub J-P, Noura H, Salman O, Chehab A (2020) Security analysis of drones systems: Attacks, limitations, and recommendations. Internet of Things 11:100218. https://doi.org/10.1016/j.iot.2020.100218
26. Werth AW, Morris TH Prototyping PLCs and IoT Devices in an HVAC Virtual Testbed to Study Impacts of Cyberattacks. In: Proceedings of Fifth International Congress on Information and Communication Technology: ICICT 2020, London, Volume 1, 2021. Springer, pp 612–623
27. Rustagi A, Manchanda C, Sharma N IoE: A boon & threat to the mankind. In: 2020 IEEE 9th International Conference on Communication Systems and Network Technologies (CSNT), 2020. IEEE, pp 114–119. https://doi.org/10.1109/CSNT48778.2020.9115748
28. Afzaliseresht N, Miao Y, Michalska S, Liu Q, Wang H (2020) From logs to stories: human-centred data mining for cyber threat intelligence. IEEE Access 8:19089–19099. https://doi.org/10.1109/ACCESS.2020.2966760
29. Samtani S, Abate M, Benjamin V, Li W (2020) Cybersecurity as an industry: A cyber threat intelligence perspective. The Palgrave Handbook of International Cybercrime and Cyberdeviance: 135–154
30. Soe YN, Feng Y, Santosa PI, Hartanto R, Sakurai K (2020) Towards a lightweight detection system for cyber attacks in the IoT environment using corresponding features. Electronics 9 (1):144. https://doi.org/10.3390/electronics9010144
31. Moldovyan D, Moldovyan A, Moldovyan N (2021) A new design of the signature schemes based on the hidden discrete logarithm problem. Quasigroups and Related Systems 29 (1):97–106

Chapter 4
Cyber-Resilience, Principles, and Practices

Hilary Meagher and Lubna Luxmi Dhirani

4.1 Introduction

In a world where sophisticated technologies have fully reformed ways of communication, healthcare, manufacturing, etc., has increased the need for securing these digitally transformed environments as well [1]. As per recent statistics, 5.3 billion people across the world use the Internet; public cloud usage has increased, and cloud spending has touched the $490.3 billion mark [2]. The Industrial Control Systems Operational Technology (ICS-OT) cyber-attacks surged by 60%, and 1,300 ICS-specific vulnerabilities were identified [3], the majority of which had high to critical severity ratings. It is anticipated that by 2030, more than 29 billion IoT devices will be used for industrial and commercial use; cloud dependencies on Software-as-a-Service (SaaS), Platform-as-a-Service (PaaS), and Infrastructure-as-a-Service (IaaS) and usage will grow beyond 200 zettabyte [4]. Internet of Everything (IoE) is one of the examples that emerged with new integrated technologies and communication systems. This is merely an example of the data-driven digital economy and markets we are heading toward. In the past few years, manufacturing industries have been the most exploited and cyber-extorted environments by malicious actors with the intention of gaining financial advantages, espionage, intellectual property theft, etc. These threats would potentially escalate with the use of

H. Meagher
Department of Electronic & Computer Engineering, University of Limerick (UL), Limerick, Ireland

L. L. Dhirani (✉)
Department of Electronic & Computer Engineering, University of Limerick (UL), Limerick, Ireland

Confirm—SFI Smart Manufacturing Centre, Limerick, Ireland
e-mail: lubna.luxmi@ul.ie

K. Naseer Qureshi et al. (eds.), *Cybersecurity Vigilance and Security Engineering of Internet of Everything*, Internet of Things,
https://doi.org/10.1007/978-3-031-45162-1_4

emerging technologies like IoE, as standards to control and mitigate these threats have not been fully developed. As the cyber-threat landscape is constantly changing and new threats are coming to the surface, in such a scenario, a systematic cybersecurity framework is required to identify, assess, align, mitigate, and build cyber-resilience in the environment.

To protect the manufacturing environment from increasing vulnerabilities and threats, there is a pressing need for strong cyber-policies, laws, and controls especially for IoE smart industry networks. The European Union has developed new regulations and frameworks (i.e., EU Cybersecurity Act 2022, EU Cybersecurity Strategy, EU Cyber Resilience Act 2022, EU Digital Markets Act, EU Digital Fairness Act, Network and Information Security 2-Directive (NIS2-D), General Data Protection Regulation (GDPR), Cyber Diplomacy Toolbox 2022, and 5G Toolbox [5–7]) for mitigating cyber and geopolitical risks. The war between Russia and Ukraine has demonstrated that cyber-wars could cause more damage than the ones fought at the line of control at the border. The goal of a cyber-war is to use digital technologies for hacking and targeting military, government networks, and critical infrastructure, such as power grids or transportation systems, disrupting communications, jeopardizing the availability of services (i.e., triggering economic and operational disruption), and affecting human lives [8]. These cyber-attacks are usually carried out by infiltrating the command and control (C&C), installing malware, or launching distributed denial-of-service (DDoS) type of attacks [8].

A report in [9] states that more than 45% of industries have insufficient security measures. This shows the majority of industries have no effective incident response or mitigation process in place. There could be a number of reasons to blame such negligence (i.e., lack of skilled staff, policies, security controls, measures, standards, compliance, etc.). Considering the cost related to escalating cybercrime and attacks (i.e., man-in-the-middle, crypto-jacking, phishing, third-party, software and hardware vulnerabilities, ransomware, etc.) happening at different levels in the manufacturing supply chain, if these risks are not neutralized and mitigated, they could cause massive safety, security, operational downtime, and financial consequences. Stolen intellectual property (IP) can cost manufacturing industries more than 3,000 million dollars, reputational damage, legal and litigation costs, and lost customers.

As per the new European Regulations, building cyber-resilient within infrastructures is now mandatory for industries operating in the European jurisdiction. Some of the challenges to reaching the cyber-resilient goals are as follows: (i) cross-domain interoperable standards for emerging technologies; (ii) gap analysis; (iii) regulatory and legal compliance; (iv) enforcing policies, auditing, and having an incident response plan; and (v) easing trade across Europe [9]. To secure an environment, it is essential to fully understand the technological infrastructure, operations, dependencies, resources, and flow of data.

This chapter provides a roadmap for building cyber-resilience within an industry by the following:

(i) Identifying, assessing, and aligning cybersecurity standards across the manufacturing plant
(ii) Enabling cross-platform standard alignment

Fig. 4.1 Building cyber-resilience using CYBER INTEL framework

(iii) Designing and mapping the cybersecurity strategy with the statutory, regulatory, and contractual standards
(iv) Gap analysis and threat mitigation
(v) Enforcing strong technical, operational, and political policies
(vi) Auditing and having an incident response in place
(vii) Enabling a trust-based manufacturing environment, easing international trade

The chapter is categorized in the following sections as shown in Fig. 4.1.

Section 4.2 introduces the authors' designed cybersecurity framework for building cyber-resilience. The section is further divided into ten parts that demonstrate mapping with the statutory, legal, regulatory, and contractual standards and controls. It also provides implementation using a use-case example. Section 4.3 provides a reflective summary of the chapter and future directions.

4.2 Building Cyber-Resilience in Industry Using CYBER INTEL

As technology continues to advance, the threat of cyber-war is becoming increasingly real. Critical infrastructures, industries, militaries, and governments around the world are investing heavily in cybersecurity, developing tools and strategies to defend against cyber-attacks. This chapter introduces an authors' designed "CYBER INTEL (CYBERsecurIty staNdards, risk assessment, Threat Intelligence, Legal, and rEgulatory) framework that aligns cyber-laws and regulations, together with compliance standards and frameworks, auditing, and controls required to protect industries from the impacts of cyber-attacks. The framework provides a roadmap for selecting and employing appropriate cybersecurity standards and baseline security metrics and defines strategies for risk management and compliance with cybersecurity frameworks. The framework presents guidance for aligning related cybersecurity regulations and laws in protecting critical assets. It also provides oversight on how to build cyber-resilience in an industry. It touches on incident response planning and highlights the need for security awareness training, risk assessments, and auditing to ensure that companies are compliant with defined controls, standards, laws, and regulations that have been implemented to protect the company from cyber-crime.

4.2.1 Traction Plc. (Selected Use-Case)

For demonstrating a working example, the authors chose to use a fictional Manufacturing plant model to base the use-case on. The use-case (Traction Plc.) is derived from the author's exposure and experience gained working in the Supply Chain Manufacturing sector over the past 15+ years and disseminates valuable insights. There were data protection regulations and security issues with sharing information related to real manufacturing environment, so the authors felt that a use-case would be an appropriate choice.

Traction Plc. is a manufacturing company based in Ireland. It has three manufacturing plants located in Galway, Dublin, and Cork all of which are connected via a company Wide Area Network (WAN) which is managed by a third-party service provider [10]. Each site has a Local Area Network (LAN) and share a common enterprise domain. The enterprise ERP solution is hosted at the head quarter plant in Dublin, critical data used in this solution is encrypted at rest. Connections are managed via web services, and these are encrypted in motion using Secure Socket Layer (SSL) certs.

The Industrial Control System (ICS) network and supporting services have been segregated in line with the ISA-95 Purdue model. Each plant has its own separate physical and logical network with a common Manufacturing domain across all three. Site network perimeter firewalls are configured with content inspection enabled. Network switches are configured with Network Access Control (NAC), port security and Dynamic Host Configuration Protocol (DHCP) guard to prevent unauthorized access and unauthorized DHCP servers on the network [11]. Switch configurations are backed up using SolarWinds Network Configuration Manager to ensure recoverability. Communications into and out of the ICS network are protected by a demilitarized zone (DMZ) with firewalls. Each plant has between 5 and 10 process areas, each with its own Programmable Logic Controller (PLC) controller and a fieldbus network with various sensors and actuators. PLC firmware is updated on a regular basis in line with vendor recommendations. There is a supervisory level SCADA solution local to each plant which is used to control the processes. This interacts directly with Level 0 devices via local HMI's. ICS traffic is isolated using dedicated virtual local area networks (vLAN) per process area which are configured to limit inter-vLAN communications.

Remote access for maintenance and support is managed via a secured solution which allows connectivity through a virtual private network (VPN) connection and limits access to defined IP's and ports based on predefined requirements and multi-factor authentication. A remote desktop services (RDS) server is hosted in the DMZ and vendors with appropriate authorized access can jump to an Engineering workstation at Level 3 which has several OT applications installed to allow Engineering teams configure and maintain Level 0 and Level 1 devices. All sites have a local industrial backup and recovery solution for backing up PLC code and firmware, vendor human machine interface's (HMI) and managed Industrial switches. Servers are backed up nightly by a site backup and recovery solution with encrypted backups. Anti-virus, end point protection and regular operating system (OS) patching is in place on all Wintel devices with emergency patching for high-risk vulnerabilities catered for with an out-of-band patching cycle. Enterprise and ICS applications are kept patched up to date in line with vendor recommendations.

4.2.2 Cyber-Threat Landscape

According to a 2019 study conducted by Forrestor Consulting on behalf of Armis, "*66% of manufacturing firms have encountered an IoT-related security incident*" [12]. Major impacts from a cyber-attack on manufacturing companies include data breach, loss of intellectual property, disruption and downtime leading to financial loss and reputational damage. Majority of the ICS systems are not based on the security by design principles. This flaw would allow broader gaps within the environment, exploiting it to a broader threat landscape. Traction Plc is potentially vulnerable to threats related to legacy equipment, operating systems and software vulnerabilities, lack of network micro-segmentation and configuration issues. The attack surface stretches down to the lower levels of an ICS network where an attacker with physical access could potentially use direct access cards or chips that are plugged into a device to scan for and exploit un-remediated vulnerabilities. Maintenance interfaces with no authentication can be used to gain access and control of PLC's, opening the door for an attacker to program bad inputs into a controller to change how a process is running, or indeed the components or quantities of a recipe used by a process to manufacture a product.

4.2.3 Data Security and Risk Management

Historically, ICS systems have lacked security in their design and leading to a wider attack surface which leaves them vulnerable to attacks such as an ICS or IT Insider, common, targeted or zero-day ransomware, Industrial IoT (IIoT) pivot, vendor back door [13]. To protect against the risks mentioned in Sect. 2.2, a robust data risk management strategy is required. *"Building a mitigation and prevention strategy that centers on security, vigilance and resilience can be key toward managing risk"*[14]. A starting point to building out that strategy is understanding the technology and solutions landscape that needs to be protected. This, together with the data stored and processed in the environment, are critical to business processes and therefore need security and appropriate controls in place to protect them. Consideration should be given to the three common blocks of security which comprise the security (confidentiality, integrity and availability (CIA triad)) when designing any data risk management strategy. Use of an industry standard framework such as the NIST Risk Management Framework (RMF) is a good foundation and can be tailored to suit business needs [15].

For Traction Plc., the manufacturing company described in the use-case outlined above, a data risk management plan includes the following:

(i) Defining the key data risk management roles and responsibilities to ensure the right level of accountability and ownership is in place.
(ii) Generating a detailed asset inventory of both IT and OT systems and devices such as servers, workstations, HMI's, PLC's, scan guns, printers, network switches, in-house developed applications, Commercial off-the-shelf (COTS) software applications, licenses, etc. It includes asset details such as name, hardware type, IP address, operating system, firmware etc. User access information is made available, through Active Directory where applications are configured with lightweight directory access protocol (LDAP) or similar, or through a manual list that details the local users configured on a device and their level of access. *"High value assets and high impact systems that require increased levels of protection"*[16] are identified as part of this process.
(iii) Identifying the data hosted and processed by all systems used to support Manufacturing processes including those where there is integration between the IT and OT systems. Data is classified according to a defined data classification policy (Public, Internal, Confidential, Sensitive).
(iv) Performing a cyber-risk assessment for evaluating current security posture, including both organization and systems in line with recommendations from NIST RMF. Findings are used for identifying current known cyber-risks and an actionable plan to address is built out, taking people, process, and technology impacts into account. The cyber-risk assessment is a bi-annual exercise and requires alignment with business leaders across key functions e.g. Engineering, automation, maintenance, supply chain operations, information technology, information security, etc.
(v) Choosing appropriate security and privacy controls to protect the systems based on the results of the risk assessment. Suitable controls are applied across the various business process areas to address identified risks, in alignment with the NIST Cyber Security framework and mapped to the center for internet security (CIS) Controls ensuring appropriate coverage to meet identi-

fied business risks e.g., backup and recovery, asset management, compliance management, system security, physical security, operations, disaster recovery.

(vi) Identifying resources with ownership and accountability for the controls.

(vii) Implementing the controls and ensuring process documents are complete. A detailed description of the risk and objective of the control, and the associated execution steps are required to ensure there is clarity for the assigned control and process owners who are responsible for executing and reporting the control results.

(viii) Assessing the controls after a certain period of time to ensure they are working as expected and producing the desired outcomes.

(ix) Defining and rolling out a regular process to monitor the effectiveness of the controls and ensuring they operate as expected. This should encompass mitigation activities and redesign of control processes where areas of opportunity are identified as part of regular monitoring.

4.2.4 Cyber and Data Protection Laws & Regulations

The high-level plan outlined above provides a foundational approach for protecting Traction Plc's assets. To strengthen this data risk management strategy, several cyber-laws have been defined and enacted under Irish and European Law which "*provide for various cybercrimes like hacking, phishing, electronic theft, etc. Ireland also has a multitude of laws governing data protection and privacy laws*" [30]. The core purpose of these laws is to protect critical data. For Traction Plc. which is hosted in Ireland, the following laws are applicable, and the company will need to comply with their directives:

(i) *Criminal Justice (Offences Relating to Information Systems) Act 2017*: this Act provides legislation to protect against common cyber-attacks such as hacking, malware, denial of service, identity theft/fraud as well as others. This is applicable for Traction Plc. as the likelihood that the company will experience a cyber-attack at some point is high. This Act supports legal recourse for cybercrimes [17].

(ii) *Data Protection Act 2018 (GDPR):* Traction Plc. needs to handle employee and supplier information during its standard business operations. GDPR governs how this type of personal data is controlled, processed, and stored, and governs privacy rights for an individual [18].

(iii) *EU Cyber Security Act (2019):* this Act established an EU framework for certification of digital products and services in Europe. This ensures a common cybersecurity certification approach in Europe and ultimately, improving cybersecurity in a broad range of products and services [19]. This applies to the network services provided by the third-party service provider in the use-case above, and is also applicable to any ICT products purchased by Traction Plc.

(iv) *E-Privacy (S.I No. 336/2011) European Communities (Electronic Communications Networks and Services) (Privacy and Electronic Communications) Regulations 2011:* this law regulates the way that public telecommunications network providers or services handle personal and private data. In the above use-case, the third party who are responsible for managing the company WAN would need to be compliant with this law [20].

(v) *NIS2-Directive*: this legislation *"sets the baseline for cybersecurity risk management measures and reporting obligations"* and will *"further improve the resilience and incident response capacities of both the public and private sector and the EU as a whole"* [21]. As a manufacturer based in Ireland, Traction Plc. must comply with the cybersecurity regulations enforced by this law and will need to be able to demonstrate compliance.

(vi) *EU Cyber Security Resilience Act 2022*: while not yet enacted, a proposal has been shared by the European Commission to put regulations in place to ensure that all digital products are secure by design and are kept secure throughout their lifecycle. As a consumer of digital products, Traction Plc. will need to comply with this Act and ensure that digital products are kept secure, in line with Manufacturer recommendations e.g. patched up to date, hardened [22].

4.2.5 Governance, Risk and Control – Data Protection

Based on the results of an assessment completed using the GDPR Temperature Tool [23], Traction Plc. are considered at low risk of potential exposure to GDPR sanctions (see Fig. 4.2 below). The company operates only in Ireland and does not transfer data outside the EU. Several focus areas were identified that need further review and a plan put in place to address. These include:

(i) Train employees on processing of personal data.
(ii) Complete a risk assessment on processing activities that are carried out.
(iii) Complete a Data Protection Impact Assessment (DPIA) for those processing activities which are subject to same, based on GDPR guidelines.
(iv) Confirm if Traction Plc. are required to keep records of processing activities.

4.2.6 NIST Risk Management Framework

The NIST Risk Management Framework (NIST RMF) provides a comprehensive, flexible, repeatable, and measurable 7-step process (preparing, categorizing, selecting, implementing, assessing, authorizing and monitoring the environment) [24]. This framework was considered because it aligned with Traction Plc. business and security needs.

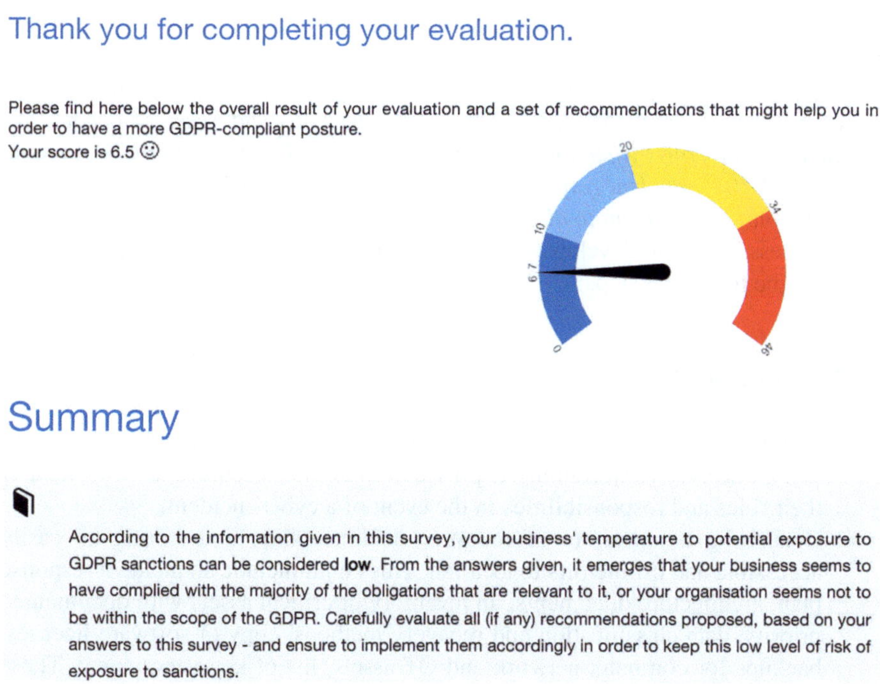

Fig. 4.2 GDPR Temperature Tool results

As the critical assets were identified in Sect. 2.3 (mentioned in i-ix), security measures related to preparing Traction Plc. for managing information security and privacy risks, categorizing critical assets based on impact analysis, selecting and applying relevant controls (NIST 800-53), evaluating the efficacy of the controls, assigned process owners who had authority for risk-based decision and continuously monitoring the controls implemented and risk matrix.

Traction Plc. had taken a proactive approach, identified the need for a risk management framework and implemented the controls before moving to the regulatory and legal frameworks. This helped the manufacturing plant in converging and smoothly aligning with the data security, statutory and regulatory controls (e-Privacy, GDPR, NIS2D, Criminal Justice (Offences Relating to Information Systems) Act 2017, National Cyber Security Strategy 2019-2024, EU Cybersecurity Act.

4.2.7 *Incident Response Planning*

One of the main components that underpin a strong cyber-resilience strategy is having a well-defined, robust incident response plan to enable companies react to and recover following a cyber-attack such as ransomware, malware, data breach etc. The plan is typically a *"written set of guidelines that instructs teams on how to prepare for, identify, respond to, and how to recover from a cyber-attack"* [25].

The NIST Computer Security Incident Handling Guide outlines a four-step process for managing incidents. It is worth noting that *"incident response is not a linear activity that starts when an incident is detected and ends with eradication and recovery. Rather, incident response is a cyclical activity, where there is continuing learning and improvement to discover how to better defend the organization"* [26]. This process is used to develop a cyber-incident response plan for Traction Plc. and includes the following steps:

4.2.7.1 Preparation

(v) Identifying key resources to form a Computer Security Incident Response Team (CSIRT) [27], training is provided to ensure resources are informed of their roles and responsibilities in the event of a cyber-incident.
(vi) Generating a repository of recovery documentation and storing an easily accessible and offline/offsite location. This could include an incident response plan, architecture documents, an inventory of critical assets with documented priority, data classification and recovery methods, copy of software licenses, backups for compute, network and OT assets, list of key vendors etc. These documents are reviewed and updated on a regular basis.
(vii) Defining clear steps to be followed if an employee notices suspicious cyber-activity.

4.2.7.2 Detection and Analysis

(i) Collect and review available data from internal and external sources to determine the type of threat per NIST guidelines (precursors and indicators).
(ii) Perform a detailed analysis to identify the vulnerabilities that have been exploited and document and prioritize post-incident actions, ensuring that an audit trail of evidence is maintained.
(iii) Prioritize the approach to handling the incident in terms of functional impact, informational impact, and recoverability.
(iv) Notify impacted parties including reporting to *"appropriate agencies, law enforcement, and any other affected parties"*.

4.2.7.3 Containment, Eradication and Recovery

(i) While the strategy for containing the incident may vary depending on the attack vector, the main objective is to stop the attack and prevent it from further damaging Traction Plc's assets and/or data. Gathering evidence and identifying the attacking hosts are key.

(ii) Following successful containment, the CSIRT team's focus moves to *"eradicating the threat, including removing malware and deleting compromised accounts"*.

(iii) Finally, a phased recovery begins *"which includes cybersecurity patches and taking steps to improve firewalls, reinstall anti-malware, restore systems from clean backups, and changing passwords across the organization"*.

4.2.7.4 Post Incident Activity

(i) A formal session is held to review the incident in depth, learn from challenges encountered during resolution, identify areas for improvement, validate all key stakeholders are part of the process and ensure incident response documentation is updated. This enables a Traction Plc to develop their *"security measures and indeed the incident handling process itself"*.

In general, it is good practice to run regular simulated cyber-incident response exercises where typical cyber-attack scenarios are played out with engagement from key stakeholders. The primary goal is to ensure cyber-recovery plans are tested, validated, and proven to demonstrate confidence in the business's defined and documented recovery procedures. Supporting this activity is regular testing of recovery from backups for compute, network, applications, OT assets, etc. A secondary output is enabling the business to develop an understanding of the average time it would take for Traction Plc. to recover from an incident. In addition, an enterprise-level business continuity plan (BCP) exists which outlines the ability to shift manufacturing capabilities from one plant to another in the event of a major crisis that results in an entire plant being destroyed. This plan also addresses supply risks from third-party suppliers and identifies a list of vetted alternates for critical materials.

If Traction Plc. were to detect a ransomware attack targeting HMIs, this would result in a high impact on operations as manufacturing would be stopped, with potentially significant financial and reputational impacts for the company. However, there is a well-defined incident response plan and a trained CSIRT team who will respond to the incident, containing and eradicating the malware. There are validated offsite copies of backups that can be used to restore HMIs, PLCs, and the Industrial switches used to enable connectivity for the process areas. As the ICS network is segmented, and traffic is contained using dedicated vLANs per process area, the risk of lateral movement is diminished. An attack on one location can be isolated to that location by virtue of the fact that the SCADA solutions are local to each plant and there is no cross-plant communication at the ICS layer.

While external shock factors such as domestic state-sponsored crime or armed conflict are unlikely to have an impact on Traction Plc., due to its location in Ireland, a resource impact cannot be ruled out due to the current macroeconomic climate. From a legal perspective, Traction Plc. is compliant with the Criminal Justice (Offences Relating to Information Systems) Act 2017 and could use legal means to gain recourse in the event of a breach.

4.3 Cybersecurity Compliance

Cybersecurity compliance refers to protecting data security, availability, and integrity. Tools such as the NIST CSF [28] enable to assess of Traction Plc's current security posture and develop a plan for managing cyber-risk. For small and medium-sized businesses like Traction Plc, the simplicity, and flexibility of the NIST CSF proves to be valuable. Traction Plc. used the Axio360 tool for implementing the NIST CSF, aligning and mapping it with NIST 800-53 (controls), IEC62443 [1], GDPR, and other required standards.

Based on the results of the Axio360 report (see Fig. 4.3 below), and understanding that developing a strong cybersecurity program is a critical but difficult task, as the global threat landscape continues to grow. Though various risks and weaknesses in the report were identified, the ones with high-risk impact required immediate attention and are mentioned below:

(i) Review the company's approach to supply chain risk management. Currently there was no defined process for assessing the cybersecurity posture of third-party partners and vendors. It was noted that consideration should be given to define a third party vendor management program including regular risk assessments, as the potential for significant impact from a breach of a third party is an unknown without this layer of governance in place.
(ii) Traction Plc. currently has limited capabilities around detection of events/anomalies and understanding their potential impact. While SolarWinds Orion is in place, it is more of a tactical monitoring tool which does not have advanced threat detection, at this stage proof of concept Security information and event management (SIEM) products such as AlienVault, Qualys, and QRadar. would benefit and increase visibility in this area.
(iii) While some work has already been done in crisis management, the report shows gaps in few processes. Engagement of external consultants who specialize in this area may be a good investment for Traction Plc.

4.4 Governance, Risk & Compliance – Audit Assurance

Earlier in the chapter we discussed a set of security and privacy controls that are implemented for protecting Traction Plc., based on the findings from a cyber-risk assessment. These controls cover multiple business processes, encompassing areas such as system security, physical security, configuration management, operations, backup and recovery, disaster recovery etc. Control techniques and test scripts were defined, control and process owners identified, and a regular monitoring process put in place to ensure compliance. These controls were based on NIST CSF standards but are also mapped across the CIS Controls.

The purpose of a cybersecurity audit is to carry out a "systematic and independent examination of an organization's cybersecurity and to ensure that the proper

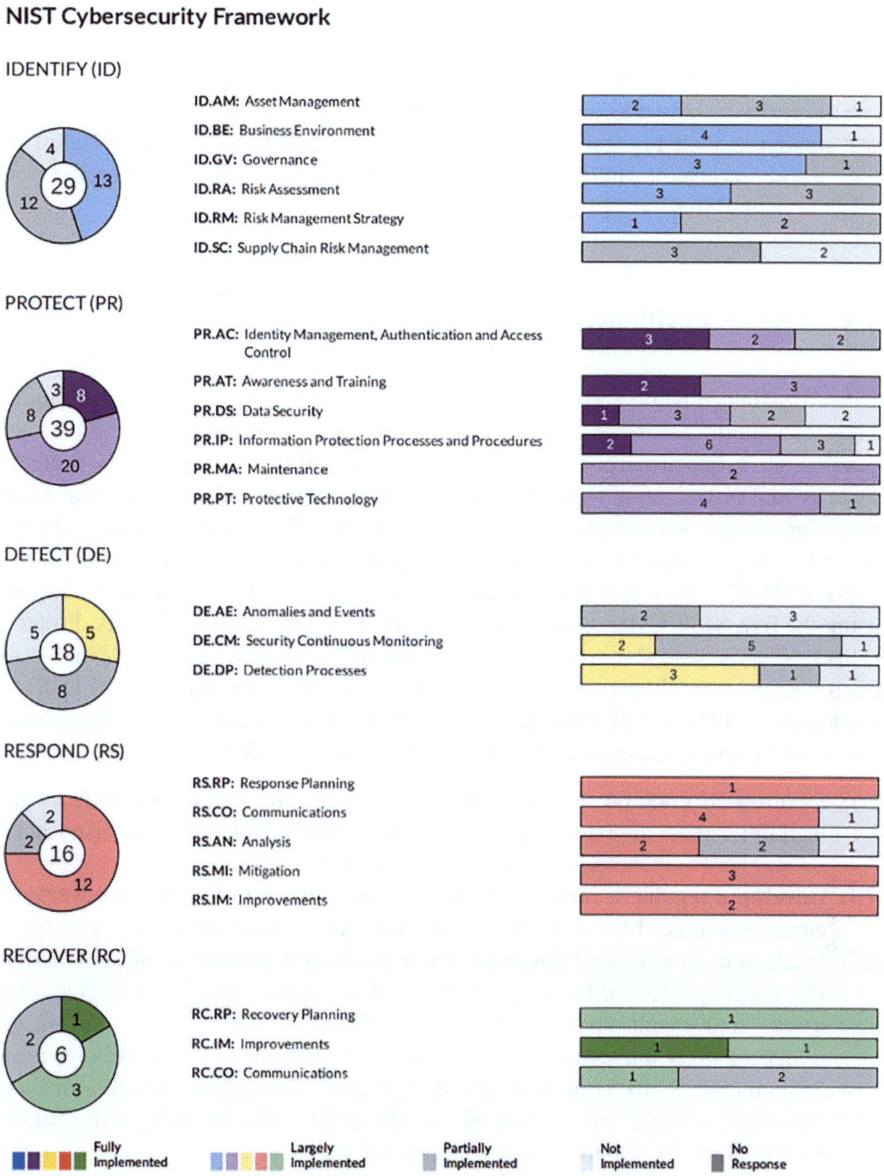

Fig. 4.3 Traction Plc's Cybersecurity posture after implementing the security standards, the gaps mentioned can be seen. These gaps identified can be easily mitigated by using the steps mentioned in Cyber INTEL framework (see 2.8 (i-iii))

security controls, policies and procedures are in place and working effectively". Engaging a third party to perform an independent audit has significant benefits, including providing assurance to the business that governance, risk, and control processes are in place and are compliant with standards and regulations. There is the added benefit of discovering potential risks or compliance issues that may exist. Audits are typically mandatory for companies to prove compliance with industry cybersecurity frameworks and laws e.g., NIST CSF, CIS, GDPR etc. There are several standards available that can be audited against, SSAE-18 and AT-101 are two such standards which can be used *"to review controls of technology Vendors and other Service Providers"* [28].

4.5 Cyber-Resilience

Cyber-Resilience is the ability to anticipate, withstand, recover from, and adapt to adverse conditions, stresses, attacks, or compromises on systems that use or are enabled by cyber-resources. It brings "business continuity, information systems security and organization resilience together" [29]. All areas discussed so far form part of an overall cyber-resilience strategy for Traction Plc. However, there is always room for improvement when it comes to cybersecurity and we can build on the NIST RMF [27] data risk management strategy discussed earlier in this chapter, using the five NIST Cyber Security Framework (CSF) functions – Identify, Protect, Detect, Respond and Recover. Note that Traction Plc. is not a FinTech, US government agency or healthcare provider and so compliance standards such as DORA, FedRAMP, e-PHI, and HIPAA. are not applicable for this use-case. Some actions that would provide the most benefit for Traction Plc. include the following:

(i) Develop information security policies to ensure roles and responsibilities are well defined for anyone with access to critical data, including safe disposal of assets (hardware, software, data) which are end of life.
(ii) Develop a regular security awareness training program for all employees, to ensure resources have a good level of awareness around suspicious activity.
(iii) Implement an advanced threat detection monitoring solution to allow Traction Plc. build up a *"baseline of expected data flows and operations for systems and users"* and enable analysis of *"detected threat events to better understand attack methods and targets"*. Both a Host-based Intrusion Detection/Prevention System (HIDS/HIPS) such as Splunk, Symantec Data Centre Security or Palo Alto Cortex XDR, and a Network-based Intrusion Detection System (NIDS) such as Snort or Splunk, should be considered.
(iv) Develop a strong incident response plan to empower business continuity in the event of a cyber-attack.
(v) Consider engaging a reputable third party to conduct annual penetration testing to validate current security posture and identify any potential vulnerabilities.

4.6 Enhanced Cybersecurity Posture Achieved Using the CYBER INTEL Framework

In this chapter, the authors selected a use-case manufacturing company and assessed its security posture using an overview of the foundational security layers, tools, and processes that were in place to protect Traction Plc's critical business assets. The authors developed an awareness of the types of cyber threats that this use-case would be susceptible to and applied a Defense-In-Depth security strategy using the CYBER INTEL framework (based on cybersecurity standards, regulatory compliance frameworks, and cyber-laws and regulations) to secure the ICS environment. Below is a brief synopsis of the areas that have been strengthened:

(i) A data risk management strategy was developed which focused on protecting critical assets and data and mitigating against cybersecurity risks. This strategy involved defining a RACI (responsible, accountable, consulter, and informed) matrix, compiling an asset inventory, identifying and classifying data, performing a cyber-risk assessment to identify current cyber-risks, building out an actionable plan, and implementing appropriate security and privacy controls, including regular monitoring.
(ii) Compliance with industry-relevant cyber-laws, e.g., Criminal Justice Act 2017 [31], NIS2-Directive [13], etc., to protect critical data and provide a path for legal recourse in the event of a breach.
(iii) Compliance with GDPR [32] with a plan to review several identified areas for further assessment and potential remediation.
(iv) Compliance with NIST CSF with a number of focus areas identified that would strengthen Traction Plc's security posture.
(v) A cyber-resilience strategy was developed to ensure that Traction Plc. can continue to operate and deliver products despite cyber-incidents.
(vi) A robust and proven incident response plan to enable Traction Plc. to react to and rapidly recover from a cyber-attack.
(vii) Cyber-auditing to provide assurance that cyber-risk processes and standards are in place and functioning effectively. Auditing will identify any potential security weaknesses that need to be strengthened and will provide assurance that Traction Plc. is compliant with the relevant cyber-laws, including GDPR, in the form of a SOC2 report.

The EU Cybersecurity Resilience Act 2022 is a proposal for a regulation that will ensure the development of digital products, such as hardware and software. The Act will also place an onus on consumers to assess and choose products that meet their security requirements and to ensure that those products are kept secured from cyber-threats for their lifecycle, in line with manufacturer recommendations e.g., hardening, secure firewall configurations, patched up to date, etc. This Act aligns with the previously mentioned cyber, legal, and regulatory standards and sets the tone for building a mature cybersecurity posture.

As a manufacturing company, Traction Plc. relies heavily on technology such as network interfaces, microcontrollers, industrial firewalls, computational resources, operating systems, etc. and will need to comply with these Acts. They will be subject to compulsory external audits to ensure compliance with standards. Based on the recommendations made in this chapter, Traction Plc. has good security and processes in place, but to ensure compliance with the Act, a risk assessment would need to be carried out and any further areas for improvement identified and addressed.

4.7 Conclusion and Future Directions

This chapter gives a high-level overview of the security posture of a use-case manufacturing company. It introduces and implements the authors' designed "CYBER INTEL" (CYBERsecurIty staNdards, risk assessment, Threat Intelligence, Legal, and rEgulatory) framework, which considers the cyber-threat landscape and common attack vectors that ICS networks face and touches on the risk and compliance frameworks and standards, together with cyber-laws that are in place to protect a company's critical assets. It provides a roadmap for developing a strong cyber-resilience strategy and also considers appropriate auditing standards in place for providing assurance over the correct implementation of appropriate security standards and compliance with required laws and regulations. While Traction Plc. has strong foundational security practices in place to address cyber-risk, there are opportunities to reinforce that position and better prepare the company for inevitable cyber-attacks and how best to respond to them to limit their impact. To ensure a return on cybersecurity investment is achieved, it is key that controls, compliance, and support for continued security awareness training are embedded into the company's culture. Cybersecurity and data protection laws that have been enacted in Ireland are pertinent in the battle against cybercrime because they have financial as well as legal consequences. Ultimately, one of the key overall takeaways from this chapter is the need for companies to develop strong and achievable disaster recovery and business continuity plans, which are regularly tested and continuously updated, to allow a company to continue operations and recover from inevitable cyber-attacks which is the core definition of cyber-resilience. Looking toward the future that relies on the digital economy, digital passports, and digital transformation, a world where everything depends on data, cybersecurity would be of utmost importance. The implications and impact of cyber-risk associated with the emerging technologies (i.e., Artificial Intelligence, Quantum Computing, 6G, etc.) used in digital infrastructures would be hard to assess as the technologies have not been fully realized yet and would be susceptible to the novel cyber-threat landscape. The standards for these emerging technologies are still under development, leaving a wide gap open for malicious exploitation. Building and sustaining cyber-resilience in critical infrastructures, industry, and the economy will be the biggest challenge of the future, and it will become crucial to develop new standards and frameworks and build essential skills across these emerging fields. Even in the future, the authors' designed CYBER

INTEL framework will stay intact and provide a roadmap for identifying, aligning, and implementing different cyber, legal, and regulatory standards for building cyber-resilience within the environment.

References

1. Dhirani LL, Armstrong E, Newe T (2021) Industrial IoT, Cyber Threats, and Standards Landscape: Evaluation and Roadmap. Sensors (Basel, Switzerland) 21 (11):3901. doi:https://doi.org/10.3390/s21113901
2. Worldwide semiconductor industry capital spending will decline by 47.9 percent to $22.9 billion in 2009 – Gartner (2010). Microelectronics International 27 (1). doi:https://doi.org/10.1108/mi.2010.21827aab.004
3. Jenkinson A (2022) US State Attacks and the Continued Oversight of Security. Ransomware and Cybercrime. CRC Press. doi:https://doi.org/10.1201/9781003278214-17
4. Dhirani LL, Newe T (2020) Hybrid Cloud SLAs for Industry 4.0: Bridging the Gap. Annals of Emerging Technologies in Computing 4 (5):41–60. doi:https://doi.org/10.33166/aetic.2020.05.003
5. Chiara PG (2022) The Cyber Resilience Act: the EU Commission's proposal for a horizontal regulation on cybersecurity for products with digital elements. International Cybersecurity Law Review 3 (2):255–272. doi:https://doi.org/10.1365/s43439-022-00067-6
6. Burri M, Zihlmann Z (2023) The EU Cyber Resilience Act – An appraisal and contextualization. EuZ – Zeitschrift für Europarecht. doi:https://doi.org/10.36862/eiz-euz015
7. Ashkenazi A (2022) Cyber Diplomacy 3.0 - "Agile Diplomacy" to Promote Security and Innovation. International Journal of Cyber Diplomacy 3:81–96. doi:https://doi.org/10.54852/ijcd.v3y202209
8. Dhirani LL, Mukhtiar N, Chowdhry BS, Newe T (2023) Ethical Dilemmas and Privacy Issues in Emerging Technologies: A Review. Sensors (Basel, Switzerland) 23 (3):1151. doi:https://doi.org/10.3390/s23031151
9. Embroker (2023) Must-Know Cyber Attack Statistics and Trends. www.embroker.com/blog/cyber-attack-statistics. 2023
10. Anwar RW, Bakhtiari M, Zainal A, Qureshi KN (2016) Wireless sensor network performance analysis and effect of blackhole and sinkhole attacks. Jurnal Teknologi 78 (4–3)
11. Kashif Naseer Qureshi AA, Raja Waseem Anwar, Shahid Nazir Bhati, and Gwanggil Jeon (2021) Fully Integrated Data Communication Framework by Using Visualization Augmented Reality for Internet of Things Networks. Big Data 9 (4):253–264. doi:https://doi.org/10.1089/big.2020.0282
12. Lan L, Bai J, Chen X (2020) Overview of Enterprise IoT Security System based on Edge Computing. Paper presented at the Proceedings of the 5th International Conference on Internet of Things, Big Data and Security,
13. Qureshi KN, Rana SS, Ahmed A, Jeon G (2020) A novel and secure attacks detection framework for smart cities industrial internet of things. Sustainable Cities and Society 61:102343. doi:https://doi.org/10.1016/j.scs.2020.102343
14. Lohrmann D, Tan S (2021) Cyber Mayday and the Day After: A Leader's Guide to Preparing, Managing, and Recovering from Inevitable Business Disruptions. John Wiley & Sons,
15. Salbert A (2019) Compatibility of Polish Law with EU Law Concerning the Use of Electronic Communications Means for Direct Marketing Purposes. Yearbook of Antitrust and Regulatory Studies (YARS) 12 (19):53–73
16. NIST Privacy Framework: A Tool for Improving Privacy Through Enterprise Risk Management, Version 1.0 (2022). National Institute of Standards and Technology. doi:https://doi.org/10.6028/nist.cswp.10

17. Government and Oireachtas (2018). The Constitution of Ireland. Hart Publishing. doi:https://doi.org/10.5040/9781509903467.ch-003
18. Dove ES (2018) The EU general data protection regulation: implications for international scientific research in the digital era. Journal of Law, Medicine & Ethics 46 (4):1013–1030
19. Papakonstantinou V (2022) Cybersecurity as praxis and as a state: The EU law path towards acknowledgement of a new right to cybersecurity? Computer Law & Security Review 44:105653. doi:https://doi.org/10.1016/j.clsr.2022.105653
20. Harding M (2011) The Curious Incident of the Marriage Act (No. 2) 1537 and the Irish Statute Book. SSRN Electronic Journal. doi:https://doi.org/10.2139/ssrn.1742858
21. Avramidou M, Biasin E, Kamenjasevic E, Kun E, Nisevic M Cybersecurity and the NIS2 Directive: regulatory aspects and sectoral perspectives. In: Consolidated Proceedings of the Second ECSCI Workshop on Critical Infrastructure Protection and Resilience, 2023. Steinbeis-Edition,
22. Chiara PG (2022) The Cyber Resilience Act: the EU Commission's proposal for a horizontal regulation on cybersecurity for products with digital elements: An introduction. International Cybersecurity Law Review:1–18
23. Piras L, Al-Obeidallah MG, Pavlidis M, Mouratidis H, Tsohou A, Magkos E, Praitano A (2021) A data scope management service to support privacy by design and GDPR compliance. Journal of Data Intelligence 2 (2):136–165. doi:https://doi.org/10.26421/jdi2.2-3
24. Maclean D (2017) The NIST risk management framework: Problems and recommendations. Cyber Security: A Peer-Reviewed Journal 1 (3):207–217
25. The Top 20 Cyberattacks on Industrial Control Systems (White Paper). (Ginter A (2019)). waterfall-security.com/20-attacks/. Accessed Accessed 7 Mar 2023
26. Cichonski P, Millar T, Grance T, Scarfone K (2012) Computer Security Incident Handling Guide: Recommendations of the National Institute of Standards and Technology. National Institute of Standards and Technology. doi:https://doi.org/10.6028/nist.sp.800-61r2
27. How to Design a Cyber Incident Response Plan for Your Business Embroker. (2022). www.embroker.com/blog/cyber-incident-response-plan. Accessed 18 Mar 2023
28. Davis RE (2021) Cyber Security Governance Audit. Auditing Information and Cyber Security Governance. CRC Press. doi:https://doi.org/10.1201/9781003099673-8
29. Ross R, Pillitteri V, Graubart R, Bodeau D, McQuaid R (2021) Developing Cyber-Resilient Systems. National Institute of Standards and Technology. doi:https://doi.org/10.6028/nist.sp.800-160v2r1

Chapter 5
Future Cybersecurity Challenges for IoE Networks

Saleem Iqbal, Saqib Majeed, and Syed Amad Hussain Shah

5.1 Overview

The perpetual expansion and interlinking of diverse Internet of Everything (IoE) devices/systems have inherent risks, and accordingly, susceptibilities grow exponentially. This probes into subsequent key aspects of apprehension, which includes the Introduction, working architecture, application, the security architecture of the Internet of Things (IoT) as compared to IoE, generic challenges of IoE network, cybersecurity challenges, network vulnerabilities, future threads, and the last which concludes the chapter. Throughout this chapter, paramount importance is given to the imperative for uninterrupted exploration, advancement, and cooperation among relevant entities to tackle these forthcoming trials.

5.2 Introduction

Internet of Everything (IoE) refers to the interconnected network of devices, objects, and people that are able to communicate and exchange data with one another through the Internet [1]. IoE is an extension of the Internet of Things (IoT) and encompasses a broader range of connected technologies, including sensors, wearables, machines, vehicles, buildings, and cities [2]. The IoE concept suggests

S. Iqbal (✉)
Department of Computer Science, Allama Iqbal Open University, Islamabad, Pakistan
e-mail: ssaleem.iqbal@aiou.edu.pk

S. Majeed · S. A. H. Shah
University Institute of Information Technology, PMAS-Arid Agriculture University, Rawalpindi, Pakistan

K. Naseer Qureshi et al. (eds.), *Cybersecurity Vigilance and Security Engineering of Internet of Everything*, Internet of Things,
https://doi.org/10.1007/978-3-031-45162-1_5

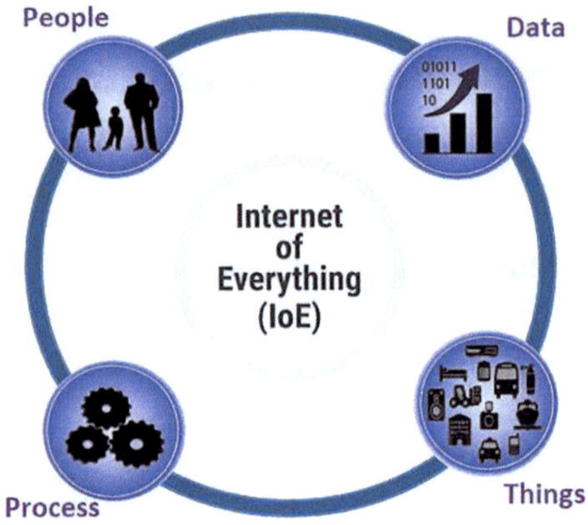

Fig. 5.1 Internet of everything

that every object, whether it is a device, machine, or even a human being, can be given a unique identifier and connected to the Internet [3]. This enables the object to exchange data and communicate with other connected objects, as well as with humans who are able to access and control these objects remotely. Figure 5.1 shows the IoE network overview.

The IoE is made possible by advances in technology, including the increasing availability of sensors and smart devices, the advancement of machine learning and artificial intelligence, and the growth of cloud computing and big data analytics. These innovative technologies empower the gathering and examination of extensive volumes of information, facilitating advancements in effectiveness, productivity, and overall well-being. The possibilities of the IoE are vast and wide-ranging, encompassing diverse industries and sectors [4]. In the realm of healthcare, IoE has the potential to enable remote patient monitoring and personalized healthcare delivery. Likewise, within the manufacturing domain, IoE can optimize production processes and minimize operational disruptions. In the context of smart cities, IoE offers solutions to manage traffic flow and curtail energy consumption. Moreover, in agriculture, IoE presents opportunities to enhance crop yields and minimize wastage. The IoE represents a significant shift in the way we think about technology and the role it plays in our lives. Through forging deeper connections across the globe, the IoE possesses the transformative capability to reshape our lifestyles, occupations, and engagement with the surrounding environment, transcending conventional boundaries.

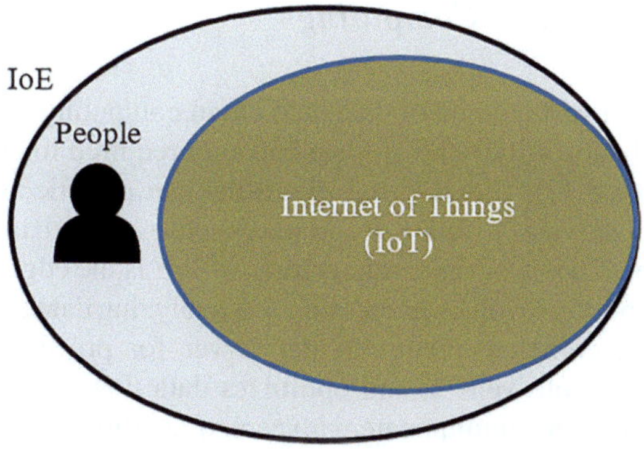

Fig. 5.2 IoE working basic architecture

5.3 IoE Working Architecture

The architecture of the IoE is constantly evolving as new technologies and applications are developed. However, there are several additional key components that are generally included in IoE architectures [5]. Figure 5.2 shows the IoE network architecture.

5.3.1 Sensors and Devices

Sensors and devices serve as the tangible entities responsible for gathering information and engaging with the world around us. They encompass a broad spectrum of physical objects, ranging from wearable technology and intelligent devices to environmental sensors, all working harmoniously to capture and interact with data.

5.3.2 Connectivity

Connectivity forms the lifeline for IoE devices, facilitating seamless communication and exchange of data [6]. To establish this vital link, diverse methods come into play, including Wi-Fi, cellular networks, and Bluetooth, among others. These avenues of connectivity enable IoE devices to stay interconnected with the Internet, enabling uninterrupted data transfer and communication.

5.3.3 Cloud and Edge Computing

The IoE engenders a vast volume of data, and cloud computing furnishes the essential computational capabilities and storage capacity required to effectively manage and process this data [7]. Cloud-based platforms can provide analytics, machine learning, and other services to analyze and make sense of the data generated by IoE devices. One of the latest working architectures of IoE is the edge computing architecture. Edge computing involves processing and analyzing data closer to the source, instead of sending complete record to the server for processing. This reduces latency, improves responsiveness, and optimizes data flow. In an edge computing architecture, devices can communicate with each other directly or with a local edge server, which performs the necessary processing and analytics.

5.3.4 Big Data Analytics

The huge amount of records created by IoE devices can be analyzed using big data analytics techniques [8] to extract insights and make informed decisions.

5.3.5 Artificial Intelligence and Machine Learning

Artificial intelligence (AI) and machine learning (ML) technologies have the potential to develop innovative solutions that offer predictive models based on the data generated by IoE devices [9]. These models can help to optimize processes, detect anomalies, and make automated decisions.

5.3.6 Security

Security is a critical component of IoE architecture, as the interconnectedness of devices and systems can create new vulnerabilities [10]. Security measures such as encryption, access controls, and threat detection need to be built into IoE systems to protect against cyber-threats.

5.4 Applications of IoE

The applications of IoE are diverse and far-reaching as shown in Fig. 5.3, spanning many different industries and sectors. Some of the applications of IoE include the following:

Smart cities leverage the potential of IoE to cultivate urban environments that are not only highly efficient but also sustainable. This is achieved by optimizing the flow of traffic, curbing energy consumption, and enhancing essential public services such as transportation and waste management [11]. The IoE acts as a catalyst, enabling cities to evolve into intelligent ecosystems that promote the well-being of their residents while minimizing their environmental footprint.

Another area of IoE is healthcare in which it is used to monitor patients remotely, deliver personalized care, and improve the efficiency of healthcare services. In manufacturing IoE also plays a pivotal role in the manufacturing sector by driving the optimization of production processes, mitigating downtime, and elevating the overall quality of products. By harnessing the power of IoE, manufacturers can fine-tune their operations, streamline workflows, and achieve heightened levels of efficiency. This transformative technology empowers them to minimize disruptions and maximize productivity, ultimately leading to enhanced product quality that meets or exceeds customer expectations.

Also the utilization of IoE in cultivation brings about significant advancements in agriculture, enabling the optimization of crop yields, the reduction of waste, and the enhancement of resource management. This groundbreaking technology revolutionizes the way farmers operate by providing them with valuable insights and tools to make informed decisions. By leveraging IoE, agricultural practices can be fine-tuned, resulting in increased productivity, minimized wastage, and more efficient allocation of vital resources. In transportation IoE is used to improve traffic management, optimize logistics, and enhance the safety and efficiency of transportation systems [12]. IoE in energy and utilities is used to

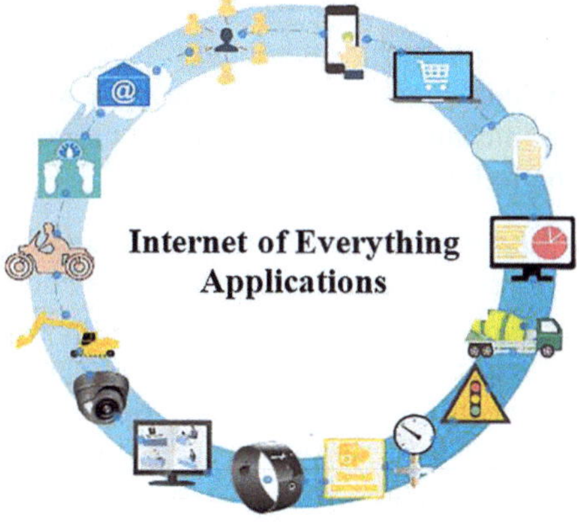

Fig. 5.3 IoE applications

monitor energy consumption, optimize energy distribution, and improve the efficiency of utilities. In retail section IoE can be used to improve the customer experience, optimize inventory management, and enhance supply chain efficiency [13]. In construction and infrastructure, IoE is used to optimize construction processes, improve the safety of construction sites, and enhance the efficiency of infrastructure management. In sports and entertainment sector, IoE can be used to enhance the fan experience, optimize player performance, and improve stadium and arena operations. By creating a more interconnected world, IoE can improve efficiency, productivity, and quality of life while also creating new opportunities for innovation and growth.

5.5 Security Architecture of IoT as Compared to IoE

The security architecture of the IoT and the IoE shares many similarities, but there are also some differences. Both IoT and IoE devices [14] are related to the Internet and are susceptible to cyber-threats, such as hacking, malware, and data breaches [14]. Therefore, security is a critical component of both architectures.

The security architecture of IoT typically includes the following components:

- *Device security*: IoT devices need to be secured through measures such as strong passwords, encryption, and regular software updates to prevent unauthorized access [15].
- *Network security*: IoT networks need to be secured using firewalls, VPNs, and other security measures to prevent unauthorized access [16] and protect against data breaches.
- *Data security*: IoT data needs to be secured through measures such as encryption [10], access controls, and data anonymization to protect against data breaches and unauthorized access [17].
- *Cloud security*: IoT cloud services need to be secured using procedures like algorithms (encryption), availability managements, and threat detection to avoid unauthorized access and protect against data breaches.

The security architecture of IoE includes all the above components but also includes some additional ones due to the broader scope and complexity of the IoE architecture. Some of the additional components of IoE security architecture include the following:

- *Edge security*: As mentioned earlier, edge computing is a key component of IOE architecture. Edge devices [18] and servers need to be secured using measures such as firewalls [19], intrusion detection, and secure boot to prevent unauthorized access and protect against cyber-threats [20].
- *Identity management*: IoE devices and users need to be identified and authenticated using measures such as multi-factor authentication and biometrics to prevent unauthorized access.

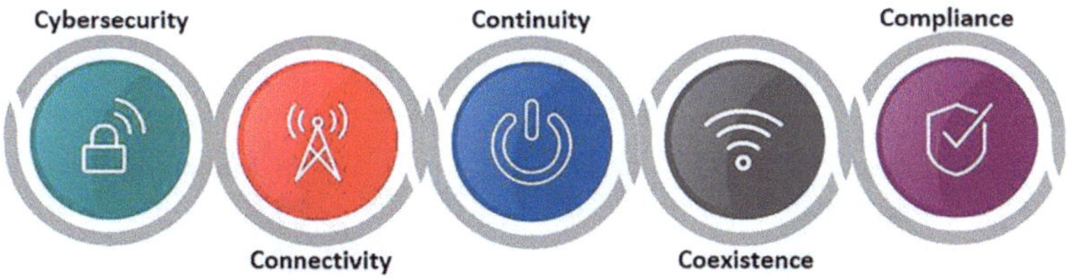

Fig. 5.4 Major challenges of IoE networks

- *Interoperability security*: IoE devices and systems need to be secured to ensure interoperability and prevent cyber-threats that may result from communication between different devices and systems [21].

The security architecture of IoE includes all the components of IoT security architecture but also includes additional components such as edge security, identity management, and interoperability security, to ensure the security and integrity of the interconnected IoE ecosystem.

5.6 Generic Challenges of IoE Network

The IoE network is a complex and interconnected ecosystem that encompasses a wide range of devices, systems, and services. As a result, there are several generic challenges that IoE networks face. Figure 5.4 shows the major challenges of IoE networks.

There are a number of challenges existing for IoE networks such as cybersecurity, interoperability, scalability, reliability, data management, power consumption, and privacy.

The challenges of IoE networks are significant, but they can be overcome through a combination of technological innovation, standards development, and best practices. As IoE networks become increasingly prevalent, addressing these challenges will become increasingly important to ensure that the benefits of IoE can be fully realized.

5.7 Cybersecurity Challenges in IoE

While IoE networks offer numerous benefits, such as increased efficiency and convenience, they also present significant cybersecurity challenges [22]. Figure 5.5 shows the cybersecurity challenges of IoE networks.

- *Exploitation of IoT devices*: IoT devices, including those incorporated within IoE networks, are frequently crafted with limited security measures, rendering them

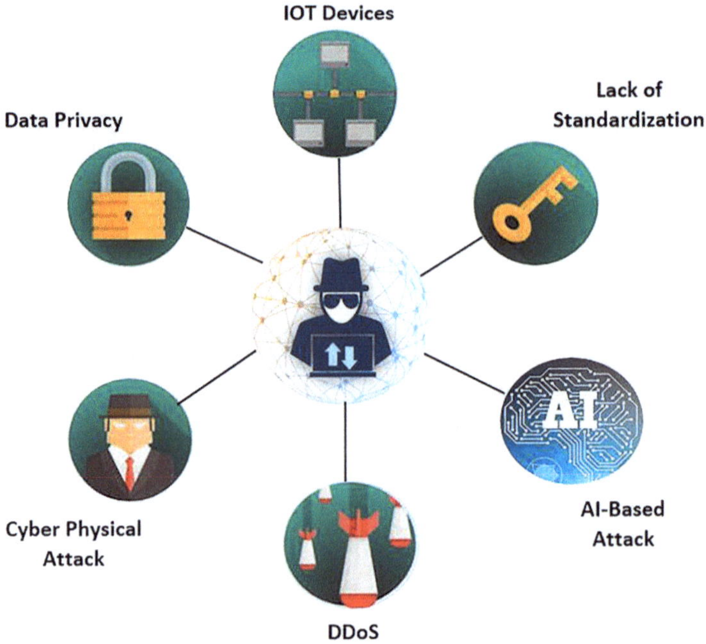

Fig. 5.5 Cybersecurity challenges in IoE networks

susceptible to exploitation. Malicious actors can leverage these devices as potential entry points to infiltrate larger networks or orchestrate extensive-scale attacks.

- *Data privacy and protection*: IoE networks generate vast amounts of data, including personal and sensitive information, which requires strict privacy and protection measures. This includes securing data in transit and at rest, implementing access controls, and ensuring compliance with regulations.
- *Attacks involving distributed denial of service (DDoS)*: IoE links often rely on cloud processing, which can be vulnerable to DDoS attacks. As IoE networks become more interconnected, the impact of a DDoS attack can be more severe, causing widespread disruption.
- *Cyber-physical attacks*: Cyber-attacks that target physical infrastructure, such as power grids, transportation systems, and healthcare facilities, pose a significant threat to IoE networks. Such attacks can result in loss of life, damage to property, and disruption of essential services.
- *Artificial intelligence (AI)-based attacks*: As IoE networks become more reliant on AI and machine learning, they also become more vulnerable to AI-based attacks [23]. Hackers can use AI algorithms to identify and exploit vulnerabilities in IoE networks, launch more sophisticated attacks, and evade detection.
- *Lack of standardization*: IoE networks often lack standardized security protocols, which can make it difficult to secure and manage them effectively. The lack

of standardization also makes it challenging to integrate different devices and systems into a cohesive network.

To address these challenges, IoE networks must prioritize cybersecurity and implement a multi-layered approach that includes strong authentication, access controls, encryption, and regular updates and patches.

5.7.1 IoE Network Vulnerabilities

The IoE network is a complex and interconnected ecosystem that can be vulnerable to various cyber-threats [24]. Here are some of the common vulnerabilities of IoE networks:

Lack of device security: Many IoE devices have limited security features, making them vulnerable to hacking, malware, and other cyber-threats. Without proper security measures, these devices can be easily compromised and used as entry points to attack the network.

Weak passwords: Weak passwords are a common vulnerability in IoE networks. Numerous devices are equipped with default passwords that often prove to be easily predictable, and unfortunately, users frequently overlook the need to modify them. This careless approach creates an opportune environment for cybercriminals to exploit and infiltrate both the devices themselves and the underlying network, gaining unauthorized access without much difficulty.

Interoperability issues: IoE networks involve devices and systems from different vendors and manufacturers, which can create interoperability issues. Devices with different security protocols may not be able to communicate securely, creating a vulnerability in the network.

Insecure communication protocols: Insecure communication protocols are a common vulnerability in IoE networks. If the communication between devices is not encrypted or authenticated, it can be intercepted and manipulated by cybercriminals.

Insufficient firmware updates: One of the significant challenges in the realm of IoE is the absence of frequent firmware updates for numerous devices. This unfortunate circumstance renders them susceptible to known security vulnerabilities, effectively leaving them exposed to potential threats. The lack of regular updates creates an enticing opportunity for cybercriminals to exploit these devices, granting them unauthorized access to the network.

Edge computing vulnerabilities: Edge computing is a precarious factor of IoE networks, but edge devices can be vulnerable to cyber-threats. The lack of security features and access controls on edge devices can create a vulnerability in the network.

Human error: Human error is a common vulnerability in IoE networks. Users may inadvertently expose sensitive information, fail to secure their devices properly, or fall for phishing scams, creating a vulnerability in the network.

IoE networks can be vulnerable to a range of cyber threats, and addressing these vulnerabilities requires a multi-layered approach that includes device security, secure communication protocols, firmware updates, and human error prevention measures. Regular security audits and risk assessments can help identify vulnerabilities and mitigate potential risks to IoE networks.

5.8 Future Threads to IoE Networks

The IoE network is constantly evolving, and with that evolution come new security threats and challenges. Here are some of the potential future threats to IoE networks:

Quantum computing attacks: This attack has the potential to break many of the encryption algorithms that currently protect IoE networks.

Autonomous attacks: These attacks could be carried out by AI-powered bots or autonomous vehicles, causing significant damage to the network.

Social engineering attacks: Social engineering attacks, such as phishing scams and social media manipulation, are becoming more sophisticated and targeted [20]. These attacks are used to gain illegal entry to IoE networks and steal sensitive data.

Supply chain attacks: As IoE networks become more complex, there is an increased risk of supply chain attacks.

Privacy concerns: In IoE networks there is a risk that sensitive data could be exposed or misused, which could lead to reputational damage and legal liabilities.

Zero-day vulnerabilities: These refer to undisclosed software flaws that have not been previously identified, making them susceptible to exploitation by malicious individuals in the cyber-realm. As IoE networks become more complex, the likelihood of zero-day vulnerabilities increases [25].

The future of IoE networks presents many security challenges and threats. Effectively countering these threats necessitates the implementation of a holistic cybersecurity strategy encompassing regular risk evaluations, leveraging threat intelligence, and adopting proactive security measures. Additionally, it is vital to remain abreast of the latest advancements in security and foster collaboration with industry peers and experts to devise robust security solutions that effectively combat evolving cyber-risks.

The scale of devices to be included in IoE is large enough and has the network of being distributed. These characteristics make IoE much relevant to blockchain-oriented solution. Therefore, in handling the cybersecurity challenges, the blockchain could be a beneficial step. As it would provide ease in tracking the individual devices, our network in terms of security lapses.

5.9 Conclusion

The forthcoming of cybersecurity in the IoE poses a multitude of hurdles that necessitate attention in order to guarantee a safeguarded and robust interconnected environment. The scale of devices to be included in IoE is large enough and has the network of being distributed. These characteristics make IoE much relevant to blockchain-oriented solution. Therefore, in handling the cybersecurity challenges, the blockchain could be a beneficial step. As it would provide ease in tracking the individual devices, our network in terms of security lapses. Perpetual exploration and advancement in the domain of IoE cybersecurity, coupled with rigorous mandates and benchmarks, have the potential to foster a more secure and durable IoE ecosystem in the times ahead.

References

1. Miraz MH, Ali M, Excell PS, Picking R (2015) A Review on Internet of Things (IoT), Internet of Everything (IoE) and Internet of Nano Things (IoNT). doi:https://doi.org/10.1109/ITechA.2015.7317398
2. Taxonomy K-b (2021) Internet of Everything (IoE) Taxonomies: A Survey and a Novel Knowledge-Based Taxonomy.1–35. doi:https://doi.org/10.3390/s21020568
3. Tyagi AK, Nair MM (2020) Internet of Everything (IoE) and Internet of Things (IoTs): Threat Analyses , Possible Opportunities for Future. 15:153–177
4. Science C FUTURE OF INTERNET OF EVERYTHING (IOE).
5. Shahzad Y, Javed H, Farman H, Ahmad J, Jan B, Zubair M (2020) Internet of energy: Opportunities, applications, architectures and challenges in smart industries. Computers & Electrical Engineering 86:106739. doi:https://doi.org/10.1016/j.compeleceng.2020.106739
6. Raj A, Prakash S (2021) Internet of Everything: A survey based on Architecture , Issues and Challenges. 2018 5th IEEE Uttar Pradesh Section International Conference on Electrical, Electronics and Computer Engineering (UPCON) (January):1–6. doi:https://doi.org/10.1109/UPCON.2018.8596923
7. Baccarelli E, Naranjo PGV, Scarpiniti M, Shojafar M, Abawajy JH (2017) Fog of Everything: Energy-Efficient Networked Computing Architectures, Research Challenges , and a Case Study. IEEE Access 5:9882–9910. doi:https://doi.org/10.1109/ACCESS.2017.2702013
8. Demirkan H (2015) Innovations with Smart Service Systems: Analytics, Big Data, Cognitive Assistance, and the Internet of Everything Innovations with Smart Service Systems: Analytics, Big Data, Cognitive. 37. doi:https://doi.org/10.17705/1CAIS.03735
9. Ryoo J, Kim S, Cho J, Kim H, Tjoa S, Derobertis CV (2017) IoE Security Threats and You. (July). doi:https://doi.org/10.1109/ICSSA.2017.28
10. Tanwar S, Popat A, Bhattacharya P (2022) A taxonomy of energy optimization techniques for smart cities: Architecture and future directions. (January 2021):1–28. doi:https://doi.org/10.1111/exsy.12703
11. Mobility SC, Services T (2019) Smart City Mobility and Transportation Services. doi:https://doi.org/10.3390/s19010001
12. Kling G, McGroarty F, Mahony MO, Ziouvelou X (2020) Heliyon Estimating the impact of the Internet of Things on productivity in Europe. 6 (May):1–7. doi:https://doi.org/10.1016/j.heliyon.2020.e03935
13. Miraz MH (2018) Internet of Nano-Things, Things and Everything: Future Growth Trends. doi:https://doi.org/10.3390/fi10080068

14. Qureshi KN, Iftikhar A (2020) 6 Contemplating Security. Security and Organization within IoT and Smart Cities:93
15. Qureshi KN, Alhudhaif A, Haider SW, Majeed S, Jeon G (2022) Secure Data Communication for Wireless Mobile Nodes in Intelligent Transportation Systems. Microprocessors and Microsystems:104501. doi:https://doi.org/10.1016/j.micpro.2022.104501
16. Yu Q, Member S, Ren J, Fu Y, Li Y Cybertwin: An Origin of Next Generation Network Architecture.1–14
17. Aliero MS, Qureshi KN, Pasha MF, Ghani I, Yauri RA (2021) Systematic Mapping Study on Energy Optimization Solutions in Smart Building Structure: Opportunities and Challenges. Wireless Personal Communications:1-37. doi:https://doi.org/10.1007/s11277-021-08316-3
18. Qureshi KN, Alhudhaif A, Azahar M, Javed IT, Jeon G (2022) A Software-Defined Network-based Intelligent Decision Support System for the Internet of Things Networks. Wireless Personal Communications. doi:https://doi.org/10.1007/s11277-022-09626-w
19. Majeed A (2016) Devising a Secure Architecture of Internet of Everything (IoE) to Avoid the Data Exploitation in Cross Culture Communications. 7 (4):328–333
20. Khan R, Member S, Member PK, Jayakody DNK A Survey on Security and Privacy of 5G Technologies: Potential Solutions, Recent Advancements and Future Directions.1–55. doi:https://doi.org/10.1109/COMST.2019.2933899
21. Qureshi KN, Shahzad L, Abdelmaboud A, Elfadil Eisa TA, Alamri B, Javed IT, Al-Dhaqm A, Crespi N (2022) A Blockchain-Based Efficient, Secure and Anonymous Conditional Privacy-Preserving and Authentication Scheme for the Internet of Vehicles. Applied Sciences 12 (1):476
22. Khalid B, Qureshi KN, Ghafoor KZ, Jeon G (2023) An improved biometric based user authentication and key agreement scheme for intelligent sensor based wireless communication. Microprocessors and Microsystems 96:104722. doi:https://doi.org/10.1016/j.micpro.2022.104722
23. Liu Y, Dai H-n, Wang Q, Shukla MK, Imran M (2020) Unmanned aerial vehicle for internet of everything: Opportunities and challenges. Computer Communications 155 (December 2019):66–83. doi:https://doi.org/10.1016/j.comcom.2020.03.017
24. Vermesan O, Friess P Building the Hyperconnected Society. CRC Press. doi:library.oapen.org/handle/20.500.12657/59721
25. Sivasankari K, Ph D, Rizvi A (2023) Modern Adaptable Zero Day Attack Detection in Network Traffic: Using Feature Identification and Tree Based Classifiers. 10 (5):466–472

Part II
Security Vigilance and Security Engineering for IoE Networks

Chapter 6
Networking and Security Architectures for IoE Networks

Fasee Ullah and Asad Ullah

6.1 Overview

For this decay, the Internet of Everything (IoE) is the underlying super network architecture and sub-class of the existing Internet of Things (IoT). This chapter broadens the reader's knowledge and piques their interest in emerging IoE networks. The chapter is broadly categorized into two main streams including securing the IoE by using advanced wire and wireless-based architecture and securing by using advanced digital image processing. The main objective of this chapter is to explore the network and security architecture for IoE networks. The importance of this research is to improve the IoE network and device security, which is necessary for the broad adoption of these technologies. Utilizing cutting-edge methods and cybersecurity precautions will assist in avoiding cyberattacks, safeguarding sensitive data, and guaranteeing the secure and dependable operation of IoE networks. We encourage more research to develop more sophisticated techniques because this study shows the potential of DIP and advanced cybersecurity measures in boosting the security of IoE devices and networks.

F. Ullah (✉)
Computer and Information Sciences Department, Universiti Teknologi Petronas, Perak Darul Ridzuan, Malaysia

A. Ullah
Department of CSE, Military College of signals, NUST, Islamabad, Pakistan
e-mail: asadullah@mcs.nust.edu.pk

K. Naseer Qureshi et al. (eds.), *Cybersecurity Vigilance and Security Engineering of Internet of Everything*, Internet of Things,
https://doi.org/10.1007/978-3-031-45162-1_6

6.2 Internet of Everything

The IoT networks are based on two terms: Internet and things. The Internet means connecting local devices by using wired or wireless networks for data sharing, and things are devices and human beings. Moreover, the IoE is an extension of IoT and was introduced by Cisco in 2013 [1]. IoE aims to make the Internet smarter by connecting complex dimensions of existing objects and artificial intelligence (AI)-based future objects. The examples of IoE applications are starting from home to commercial sectors such as transportation (railway, airlines, ships, vehicles), small and large machines, humans, and home and office appliances [2]. Thus, IoE networks connect the living things and non-living things of the world and generate heterogeneous data due to different data traffic with different generation rates.

The IoE network is a fast-developing technology, where billions of devices are connected electronically. This is a developing technology that connects billions of things, such as industrial machinery, cell phones, and sensors, to the Internet. The IoE can revolutionize several industries, including healthcare, transportation, and manufacturing, by enabling real-time equipment and process monitoring, analysis, and management [3]. These networks are based on smart and intelligent systems to facilitate the users. However, there are serious cybersecurity risks associated with this interconnectedness. The threat of cyberattacks, which can result in data theft, device malfunction, and even physical harm to people, is growing along with the number of connected gadgets. Establishing strong cybersecurity measures is crucial for the security of IoE devices and networks [4].

IoE devices and networks are vulnerable to cyberattacks because of their interconnectedness, including denial of service (DoS) attacks, port scans, malware attacks, and phishing efforts. The diversity of IoE devices and networks, the vast amount of data created by IoE devices, and the dynamic nature of network settings may make conventional cybersecurity solutions insufficient to safeguard devices and networks. As a result, there is a need for new and cutting-edge cybersecurity techniques that can adjust to the particulars for IoE networks. The IoE networks can also be secured by using cutting-edge cybersecurity methods, including intrusion detection systems (IDS), firewalls, and access control systems.

6.3 Pillars of IoE

IoE is the super extension of the IoT, and its existence is based on the four pillars of people, things, data, and process [5]. People are a critical part of IoE, which is connected to the Internet through intelligent digital devices like computers, smartwatches, and other gadgets. These devices produce data through user interaction, and users can analyze it through websites, intelligent applications, and social networking. In smart healthcare systems, the analogy of the people and their vital signs are monitored by using smart sensor nodes which forward the sensory data with the

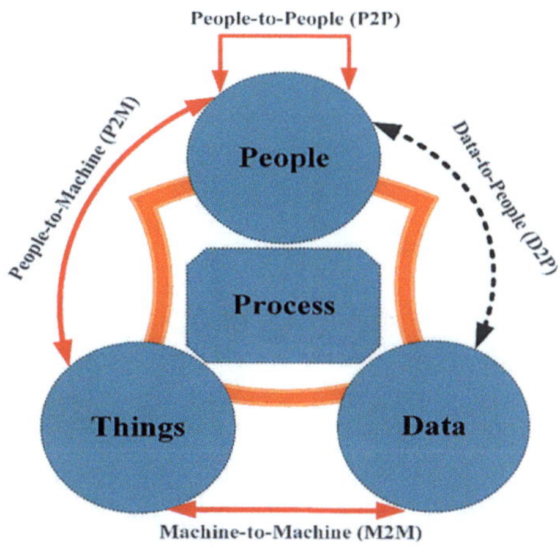

Fig. 6.1 Typical overview of Internet of Everything (IoE)

help of base station (BS) to the medical team for optimal suggestion and treatment. Moreover, the monitored vital signs of the patient's body are used to detect any abnormality in terms of low threshold or high threshold values.

Thus, people are considered to solve the problems by making different decisions to understand the choirs level of the different business groups. This whole process is situated in the people-to-people (P2P) category of the IoE environment. Furthermore, smart physical devices take instructions from people and those who interact with them through intelligent web applications. The interaction connection between people-to-machine (P2M) generated a huge volume of data to establish thoughtful and intelligent business decisions at the right time for better opportunities. The physical devices generated raw data that can be used for decision-making.

The existing IoT-based industries use AI methods with machine learning (ML) and federated learning algorithms to extract the features from gathered data and process and analyze it for better decision-making. This perfect decision process ensures customer satisfaction to grow the business networks of an organization. Figure 6.1 shows the typical overview of the main pillars of IoE networks including the processing module, which is the critical feature in gathering, analyzing, and processing data from different sources with the intervention of the people, data, and things.

6.4 Proposed Security Architecture for IoE Networks

IoE networks build relationships with social networks as a service-distributed architecture by connecting multiple devices and proposing services or protocols to other devices for different activities. In service-distributed architecture, we must design and develop a trust management system that establishes relationships between IoE

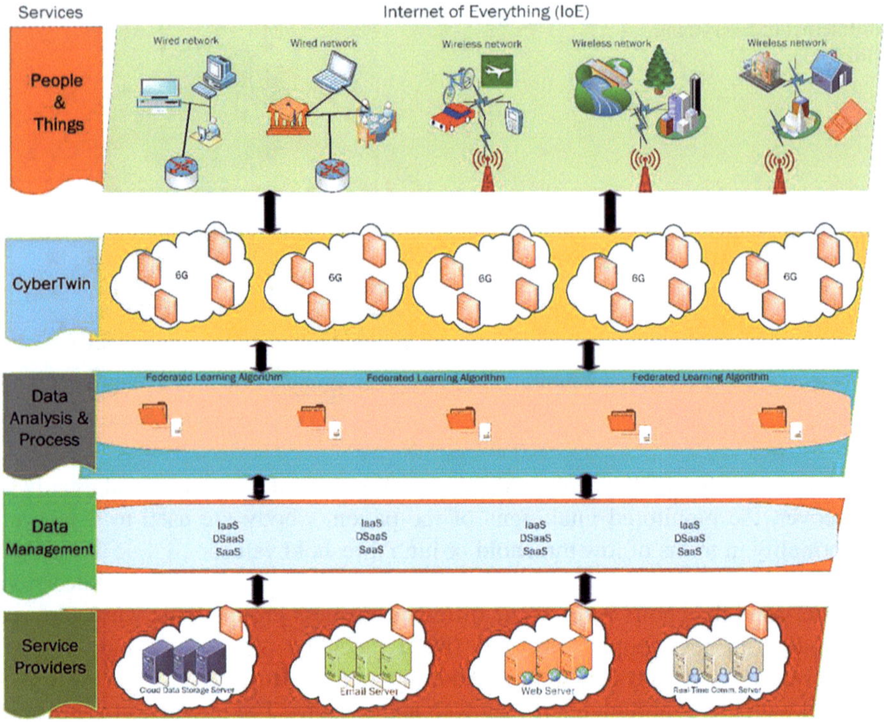

Fig. 6.2 Proposed security architecture for IoE network

devices. This section proposed a security architecture for IoE networks, comprising every living and non-living object and placing them into the people and things categories. The things or objects are computers, smartphones, and human beings which are connected by using wired and wireless-based networks through fiber optics and Wi-Fi technologies. Figure 6.2 shows the proposed security architecture for IoE networks.

Furthermore, these networks forward the data requests to the concerned servers to avail services, and the request of each object is verified for security issues using CyberTwin infrastructure. CyberTwin is the security authenticator server containing various filtering options by providing confidentiality, integrity, and availability (CIA) services. Next are the data analysis and process management services containing various machine learning algorithms. For instance, the federated learning algorithm extracts various required features in a distributed manner after clearing the security threats like spam, virus, or other attacks. Moreover, IaaS (Infrastructure as a Service), DSaaS (Data Science as a Service), and Software as a Service (SaaS) provide different types of required platforms for the required data and operations accordingly. We need different hardware and software services to allocate based on different user requests. The final part of the IoE network is the network service providers, including a cloud data storage server (CDSS), email server (ES), web server (WS), and real-time communication server (RTCS).

6.4.1 *Advanced Wire- and Wireless-Based Technologies for IoE Security Architecture*

The IoE network comprises different devices which are connected for sharing resources in the network. These networks are organized by using different technologies and standards and are further connected to other networks internally or externally for sharing the network resources. These networks are connected to different city and country networks and are called networking or inter-networking. Such examples of inter-networking are smart education, smart healthcare, smart intelligent transportation systems, smart agriculture, and smart city networks. In education institutes, the students are using Internet services by utilizing the university servers, networks, and local hosting networks. Similarly, the immigration department at the airport checks the travel history and criminal record of the passenger. Figure 6.3 shows the simple network architecture connecting different types of users using web services.

The record is extracted from centralized databases. In smart healthcare applications, a patient needs specific medicines from a particular medicine service provider, and the patient avails the online services to purchase medicines after confirmation from the concerned medical doctor. These various activities generate homogenous data traffic, and the security of the homogenous data traffic networks

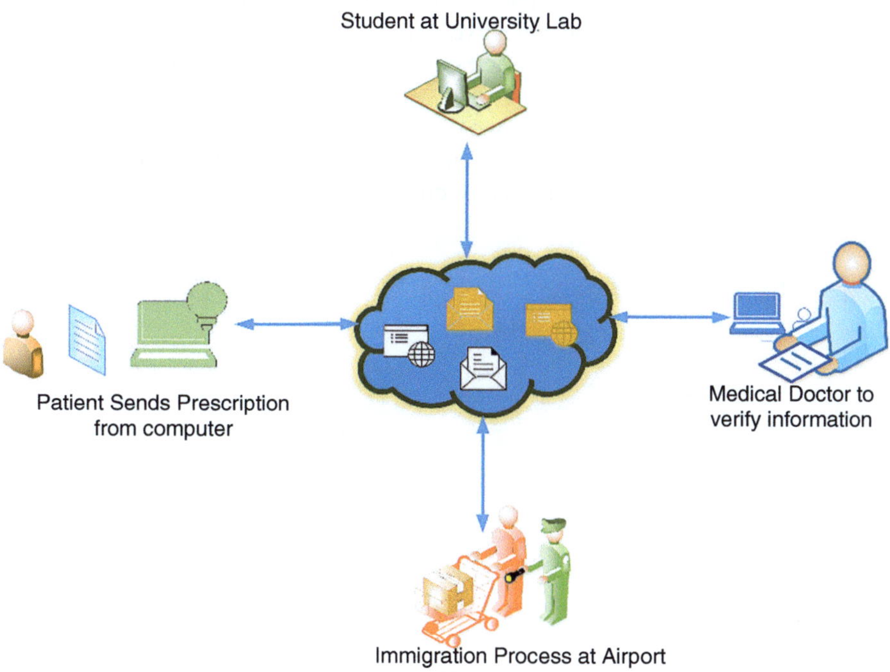

Fig. 6.3 Wired-based centralized network architecture

is comparatively easy for wireless networks to implement and handle different security threats such as DoS attacks and man-in-the-middle (MITM) attacks [6, 7].

On the other hand, wireless-based networks are used for IoE networks and devices by using different communication technologies and standards. There are different wireless communication standards used such as IEEE 802.15.4 and IEEE 802.15.6. The wireless network is the dominant technology for data transmission without fixing the wire or fixed networks. The wireless-based network architecture transmits the data in the air by using various frequencies.

The sender device generates and transmits data in the air using the registered frequency range. The receiver device (BS) receives the transmitted data and forwards it to the concerned device or network. For instance, different bio-medical sensors in healthcare are employed to monitor different vital signs of a person with three methods of installation/deployment [8, 9].

Smart healthcare technologies are used for elderly aged person home-based monitoring. The injured person who always needs continuous health monitoring or has severe health conditions patients admitted to the Intensive Care Unit (ICU). The first method is to deploy sensor nodes on the patient's body or sewed in the patient's shirt such as electrocardiography (ECG) sensors, blood pressure sensors, and temperature sensors. The second method is to implant the sensors inside the patient's body such as the endoscopy sensor to monitor and analyze different internal organs, such as kidney monitoring, liver monitoring, and taking pictures of the heart from different angles. The third method is to deploy sensors around the patient or around the patient's bed to monitor different physical activities, including sleeping duration and position, detection of the defective sitting position, and fall of the patient. Figure 6.4 shows the basic concept of the deployed IoE devices or sensor nodes for health monitoring.

Tier 1 depicted the sensor nodes or IoE devices placed inside to outside the human body for healthcare purposes. Tier 2 contains a single or multiple BSs to receive the sensory data from the body coordinator (BC), whereas BC collected

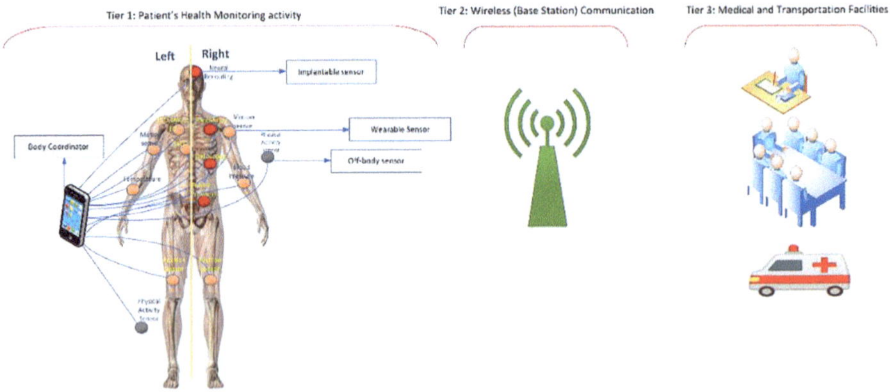

Fig. 6.4 Overview of the basic concept of the deployed BMSs network for monitoring health

sensory data from the deployed nodes, while Tier 3 contains the medical team, which analyzes the received sensory data based on the patient's medical history and suggests the optimal treatment by responding to the patient.

6.4.2 IEEE 802.15.4 Medium Access Control (MAC) Superframe Structure for Network Communication

Wireless-based communication standards are used for data communication in IoE networks. This section presents the superframe structures of IEEE 802.15.4 and IEEE 802.15.6 for handling different heterogeneous data in IoE networks. IEEE specified the IEEE 802.15.4 [10] standard for Wireless Sensor Network (WSN) connectivity. In smart healthcare systems, the sensor nodes are implants or attached to patients which track vital signs and are connected to an anatomical interface in a star topology manner [11]. The three types of patient data are regular, periodic, and emergency data. Temperature monitoring is used as normal data, whereas the glucose and blood pressure readings are taken regularly. Life-threatening vital signs are included in the emergency data. In addition, the superframe settings in IEEE 802.15.4 MAC include beacon, CAP (Conflict Progress Period), CFP (Conflict Free Period), and LPL/IP (Low Power Listening/Inactive Period). Every BMS works multiple back-off and precise channel assessments (CCA) to access the channel in contention. Furthermore, the TDMA system access mode is split into CFP slots, and the CFP time allocates the guaranteed time to transmit patient data. In contrast, the body interface allocates CFP access to BMS that received access mode in CAP times. When the sensor nodes are busy sending logical data, IP saves energy. Figure 6.5 shows the IEEE 802.15.4 MAC superframe structure.

The IEEE 802.15.4 MAC superframe configuration has the following limitations.

- The IEEE 802.15.4 superframe structure has a maximum of 16 (0–15) channels.
- During the CAP period, all deployed BMSs compete for channel access.
- Only BMSs with channel access in CAP are assigned CFP channels.

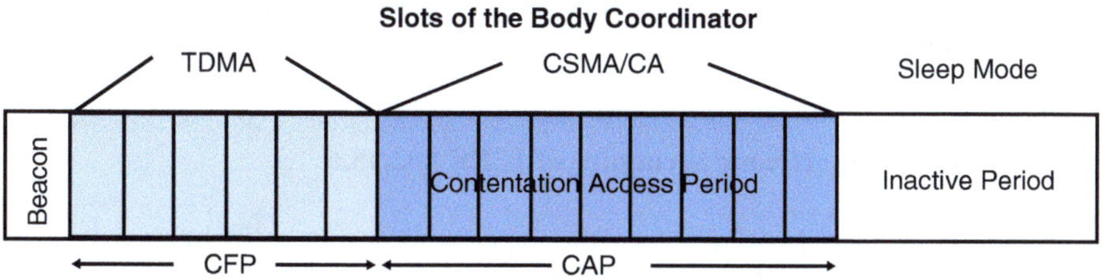

Fig. 6.5 IEEE 802.15.4 MAC superframe structure

- During channel contention, no priority-based slot is assigned to emergency data. No distinction between normal, periodic, and emergency data is made to assign the first slot based on the priority of the life-critical data.
- BMSs consume more energy and drop patient data if they exceed contention threshold values.

These constraints severely reduce the MAC superframe structure's performance in terms of higher collision; BMSs retransmit the lost data packets, causing a delay with lower reliability and a higher amount of energy consumption, which is unacceptable in an emergency.

6.4.3 IEEE 802.15.6 Medium Access Control (MAC) Superframe Structure for Network Communication

IEEE 802.15 Task Group 6 (TG6) [12] decided to develop low-power sensors for monitoring a patient's critical symptoms and the health situations of athletes in their respective sports. In 2012, the first draught version of IEEE 802.15.6 for the MAC and PHY layers was made public. IEEE 802.15.6, which divides the superframe structure into channels and beacons, is described in this draught version. Each channel is given the same time to transmit the patient's records. The IEEE 802.15.6 MAC superframe shape consists of three major modules: the MAC header, the MAC variable duration, and the Frame Check Sequences (FCS). The MAC header has 7 bytes reserved, the variable duration has 0–255 bytes reserved, and FCS has 2 bytes reserved. Furthermore, the MAC frame body is divided into three sub-headers:

- Data Freshness (one byte) to protect data from reply attacks.
- Message Integrity Code (MIC) occupies 4 bytes to authenticate the frame and maintain the frame's integrity check.
- Data payload contains data in the frame with MIC headers.

Furthermore, the IEEE 802.15.6 MAC header is divided into four sub-headers. First, the Frame Control takes up 4 bytes, distinguishes between the control and data frames, and provides an acknowledgment. The addresses of the receiver and sender sensors are specified in the second and third headers, respectively. Each sensor stores the address in 1 byte. The final header is the body coordinator header, which takes up 1 byte.

6.4.3.1 MAC Superframe Structure of IEEE 802.15.6

The beacon-enabled MAC address, the superframe structure includes a beacon, Exclusive Access Phase (EAP-I–II), Random Access Phase (RAP-I–II), Type (I–II), and CAP periods. The contention-based channel allocation policy for BMSs is

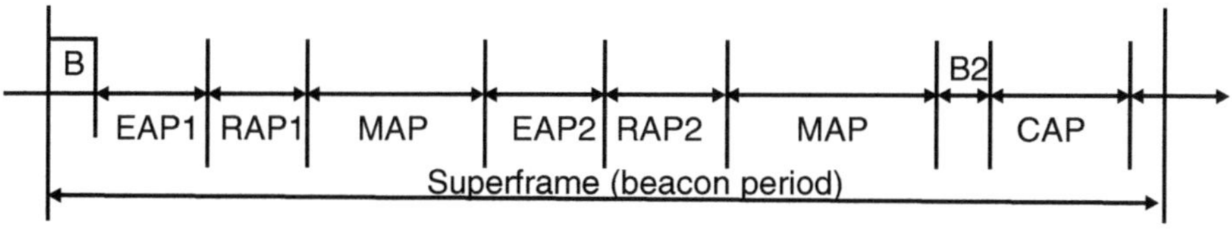

Fig. 6.6 IEEE 802.15.6 MAC superframe structure

based on CSMA/CA or slotted Aloha schedule access schemes. These access scheduling schemes are used during the EAP, RAP, and CAP periods. Type-I denotes critical data, whereas Type-II denotes non-critical data. However, the limitations of IEEE 802.15.6 MAC Superframe structure are the same as those mentioned in the IEEE 802.15.4 MAC. The non-beacon MAC Superframe structure, on the other hand, allocates the Superframe's entire channels (slots) to the Type-I or Type-II category of a patient's traffic. The disadvantage is that the body coordinator cannot directly transmit data to BMSs but must first send an activation alert signal to the recipient BMS. The non-beacon MAC also allocates slots to one type of patient's data at a time, which is unacceptable in life-critical situations. The third type is the non-beacon without Superframe structure, which uses predefined periods to transmit a patient's Type-II traffic. The slot allocation to BMSs in this Superframe is based on contention or post-contention. The restriction of predefined-based slot allocation to one type of data results in data waste (Fig. 6.6).

6.5 Data Collection, Recognition, and Processing in Multiple Environment of IoE

The IoE is an advanced concept of networks that connects multiple nature devices of different networks to collect and exchange data over wired and wireless networks. IoT has the power to receive/collect data, recognize the type of data/network, and process data for various decision-making in a central server/device. For instance, we are presenting a scenario of smart IoT-based health monitoring. There are various data traffics coming from different types of patients, and these various natures of data are called heterogeneous data because each deployed sensor node needs different frequencies or data rates to transmit the data to the designated point. For instance, the heartbeat sensor needs 1.99 kbps, the temperature sensor needs 122 bps, and ECG sensor (12 leads) needs 145 kbps [13]. Collectively, this technology receives data from different critical patients as sensory data for processing. IoE means different sensing data are collected from hundreds of thousands of small devices from multiple sources and are forwarded to the central device. The collected data is raw, and the central device needs to detect, process, and recognize by taking a specific set of actions based on the previous knowledge (using machine learning or deep learning techniques) for optimal decision.

Thus, H-IoEs assume that a patient or person uses different biomedical sensors (BMSs) whose health is monitored frequently. There are other examples like a person who is watching sports activities in the stadium, an ambulance-based patient traveling to the hospital, a person walking and exercising in parks and playgrounds, living in smart apartments, eating in a smart sensing-based restaurant, a person is traveling by a road transport, health is monitored during studies in university, a health condition is measured during sea traveling, and health monitoring of a person during working hours in public offices. Thus, the smart IoE-based sensing devices monitor different vital signs of a patient/person while a person is busy with daily life activities. The centralized server is further categorized into the database server, reasoning rules server, and main server. The database server is a simple data storage server containing the previous knowledge of a patient/person. The reasoning rules server comprises different association rules which fetch the associated data from the database server and applies certain conditions to bring the optimal decision for a patient, accordingly. However, this decision is validated by the domain expert knowledge personnel and forwarded to the central server to store it as the final decision. Figure 6.7 shows the data collection, recognition, and processing in smart healthcare IoE systems.

6.6 Diverse Technologies in IoEs

The IoE is a new technology implemented in various data communication environments. Thus, we have broadly classified IoEs as Internet of Ad hoc Network Things (IoAVTs), Internet of Smart Building Things (IoSMTs), and Internet of Underwater Things (IoUTs), with each IoE classified into more than one category of things.

6.6.1 Internet of Ad Hoc Network (IoAV)

The Internet of Ad hoc Network Things (IoAVTs) is a network where communicating objects such as laptops and smartphones are not fixed and stationary. Such devices always move from one location to another without being bound or stuck in one place or location. Furthermore, IoAV devices are outfitted with intelligent sensors and software to connect various things and transmit data gathered from the purpose-built environment over the Internet. Therefore, IoAVTs are divided into four categories: the Internet of Vehicular Ad hoc Networks (IoVAN), the Internet of Mobile Ad hoc Networks (IoMANs), the Internet of Ambulance (IoA), and the Internet of Air Traffic (IoAT).

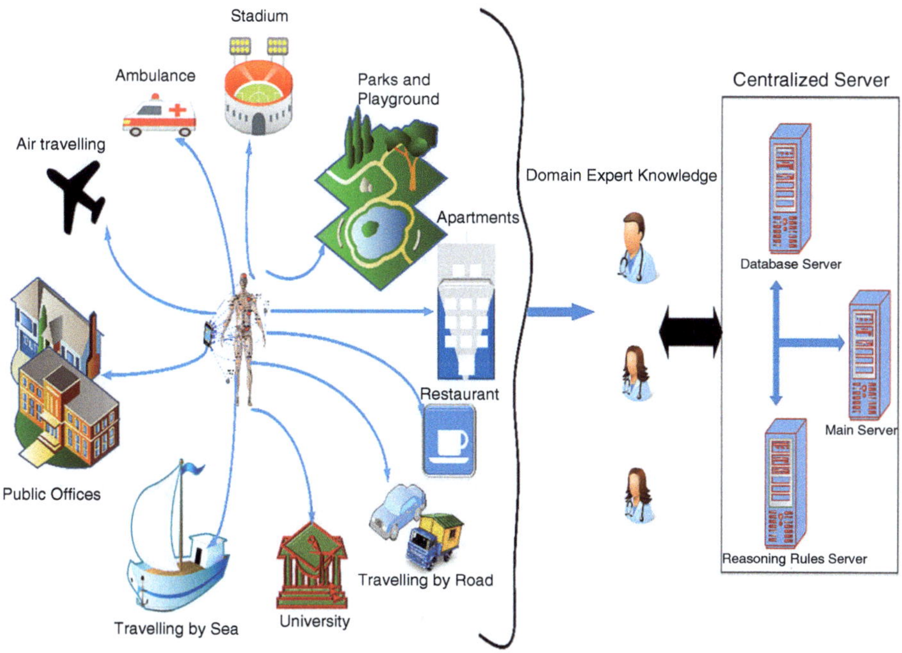

Fig. 6.7 A smart IoE support-based health monitoring system

6.6.2 Internet of Vehicular Ad Hoc Network (IoVAN)

IoVAN is made up of innovative and intelligent cars that are outfitted with advanced sensing technologies and communicate with other smart cars on the road for vehicle driving and safety. Multiple sites or roads can be outfitted with sensing devices at various points, which should transmit messages about the current state of the location/road, as well as various warning messages, via an agreed-upon standard Internet. In advanced countries, the smart car has a collision warning alert system, issuing alerts on bad driving moods such as overtaking, lousy road or weather conditions, wrong way driving, object (or vehicle) detection on the way, traffic signal and pedestrian walk violation, emergency control breaks, and notification of the hazardous location to the driver's family. Therefore, car-to-car communication should be efficient for the warnings mentioned above for the safety of the car, the people on board, and the best interests of society. Intelligent Transport System (ITS) is another name for IoVAN.

6.6.3 Internet of Mobile Ad Hoc Networks (IoMANs)

IoMAN is a network with no fixed infrastructure and consists of a collection of different mobile nodes/sensors connected in an isolated environment for data exchange. IoMANs perform network self-configuration and topology construction based on target/object detection and monitoring. All nodes use wireless channels with different radio frequencies. Furthermore, some nodes are dedicated to stationary monitoring of various activities in IoMANs, while others are regularly placed for movement/mobility monitoring of the targeted object. For example, they are tracking pandas' frequent movement and location in the forest. Thus, stationary nodes send and receive data from mobile nodes of the monitored data over the Internet to the wildlife department to ensure the animals' survival and safety. For data communication in IoMAN, there are two types of communication protocols: proactive and reactive. Proactive protocols calculate and select the best available paths from source to destination before establishing the communication channel. In contrast, the reactive protocol needs to calculate the paths in advance and instead transmits data on-demand without regard for path reliability.

6.6.4 Internet of Ambulance (IoA)

The Newport Beach Hospital and Fire Department developed the Simple Triage and Rapid Transport (START) system in 1983. The START system aims to investigate the criticalities of wounded people in mass casualties. On the side of the mass casualties, the paramedic staff arrived by assigning red, yellow, green, and black tags to the wounded to identify criticalities with severe high-risk conditions, wounded by not severe, normal, and dead people, respectively. Based on this information, the ambulance arrives on the scene and transports the injured people to nearby hospitals with available health facilities. The ambulance is equipped with all necessary first aid medication, as well as advanced wireless biomedical sensors that are installed in the ambulance to frequently monitor a patient's survival vital signs such as heart rate, respiratory rate, blood pressure, and temperature. These biomedical sensors are placed on the patient's body and are linked to a central device that collects vital sign readings and sends them to medical doctors. The readings of these vital signs are efficiently and securely transmitted to the nearest hospital's medical doctors, who are ready to treat the patient on an emergency basis.

Furthermore, the vital signs are recorded by a central device placed near the patient, and this device transmits the patient/sensory data to the medical doctors via a dedicated Internet connection. All ambulance cars exchange data on the most direct routes to hospitals while assisting patients with necessary medications. The scenario described above is known as the Internet of Ambulances (IoA). However, an efficient centralized system must be designed and developed to ensure the trust of people using IoA in mass casualties.

6.6.5 Internet of Air Traffic (IoAT)

As discussed in IoA, the Internet of Air Traffic (IoAT) is a future network that includes a mini air ambulance equipped with a wireless biomedical sensor to monitor different vital signs of patients in critical health conditions. However, there is a problem with getting patients to hospitals on time in the same city, in different cities, or moving a patient to another country due to heavy traffic or bad weather conditions. It is strongly advised in such cases to transport a patient in critical condition by air ambulance. Furthermore, the medical doctor can remotely operate the patient in an air ambulance using a satellite connection and instructions from other medical doctors. Thus, health treatment services can reduce health risks by fostering trust in IoAT services.

6.6.6 Internet of Smart Building (IoSM)

IoSM comprises advanced installed heterogeneous sensor nodes that monitor various activities inside and outside the building, for example, smoke detection inside a room, installed security cameras for surveillance, gas leakage monitoring of pipes deployed inside and outside the building, automatic door opening and closing, and electricity usage monitoring. As a result, we divide IoSM into three categories: Internet of Public Offices (IoPOs), Internet of Smart Restaurants and Hotels (IoRHs), and Internet of Smart Sports Stadiums (IoSSS). IoPO refers to federal and provincial secretariats, law and judiciary offices, post offices, electricity power distribution offices, railway station offices, military/defense offices, hospitals, weather forecasting departments, commerce and textile department, and education. These departments are outfitted with advanced nodes, sensors, actuators, and visual sensors to monitor, detect, and recognize various activities in various departments and report any suspicious activity to the appropriate authority.

The Internet of Smart Restaurants and Hotels (IoRHs) is a collection of networks linked by advanced sensing technologies that detect the mode of the client at a restaurant or hotel and provide services accordingly. Furthermore, different IoRHs are linked to provide different meal and stay services on various occasions, such as Christmas and Chinese New Year. Furthermore, the Internet of Smart Sports Stadiums (IoSSS) is a network of networks outfitted with various sensors to monitor the health of athletes during sports activities. Additionally, visual sensors are deployed around the spectator seating areas to monitor anger situations or anything suspicious and report it to the appropriate personnel. In conclusion, most future IoE networks must design an efficient network architecture without replacing the deployed hardware technologies while ensuring trust and benefits to society's citizens.

6.6.7 Internet of Underwater Things (IoUTs)

The Internet of Underwater Things (IoUTs) is a new innovative framework technology for designing and developing an intelligent communication network for underwater sensors. For data communication, IoUTs employ optical fiber and acoustic signals. Furthermore, the sea is divided into three zones. The first zone is the top surface zone of water, where a base station or wireless antenna is deployed to assist in sending and receiving data from the deep sea. The third zone is the final zone of underwater sensors deployed to monitor, detect, and identify target objects in the sea. It is assumed that the world will face a natural resource shortage after 2050, and thus the only place where the world can collect natural resources from the sea and meet their needs from deep seas, such as meat, salt, copper, gold, chemicals, oil, and gas, is the deep sea. Other advantages include detecting the earth quickly and predicting the effects of shock on the upper level of the sea or earth. Thus, the underwater deployment of sensors would detect various objects, ushering in scientific research and business revolution.

6.7 Security in IoE Networks

To safeguard the IoE networks, many cybersecurity measures are considered such as intrusion detection and prevention, access control, authentication, and encryption. To preserve data confidentiality and prevent unauthorized access, encryption method is used. To ensure that only authorized devices and users access the system, authentication entails confirming the identity of both users and devices. Access control entails granting only authorized people and devices access to resources. It entails identifying and responding to security lapses and assaults. The employment of cybersecurity precautions in protecting the IoE has been the subject of several research projects. For instance, authors in [14] proposed a security architecture for the IoE that includes encryption, authentication, and access control mechanisms. A framework for IoE intrusion detection by using ML methods is proposed in [15] and suggested employing encryption and authentication procedures to secure medical images sent over the IoE. Given the IoE's extensive use of various gadgets coupled in complex ways, it faces particular security concerns. These gadgets frequently need higher processing speed and memory due to resource-constraint environment. Traditional security solutions like firewalls and IDS systems become challenging to implement.

6.7.1 *Intrusion Detection Systems*

Discovering unauthorized access to a network or system is known as intrusion detection. Techniques for digital image processing can be used to examine network traffic and spot signs of an assault. For instance, suspicious or odd activity patterns can be found using image filtering. Feature extraction and classification might be utilized to identify particular sorts of attacks, such as denial-of-service attacks or buffer overflow attacks. Software intended to damage a computer system or network is known as malware. Malware can be found using different techniques by observing how a device's software behaves. Image filtering, for instance, can be used to spot malware-indicating patterns of behavior, such as a program that frequently accesses files or sends information to odd places. Using feature extraction and classification is possible to recognize particular kinds of malware, such as viruses, worms, or Trojan horses.

6.7.2 *Authentication*

Verifying a user's or device's identity is the process of authentication. Biometric data, such as fingerprints, facial features, or iris patterns, can be utilized to authenticate persons or devices. For instance, the contrast of an image can be improved via image filtering to make it simpler to distinguish face characteristics. Specific people can be located using feature extraction and categorization and their biometric traits. The IoE offers enormous amounts of data that may be applied to many different purposes but poses security risks. Techniques for digital image processing can be used to extract data from the massive amounts of generated image data, spot anomalies, and spot risks. The IoE can be protected using various cybersecurity techniques, such as encryption, authentication, access control, and intrusion detection. A solid approach to safeguarding the IoE can be achieved by fusing cybersecurity measures with digital image processing techniques.

6.8 Proposed DIP Architecture to Secure IoE Networks

There are serious security risks associated with this interconnectedness. The process of hiding information within an image in a way that is unnoticeable to the human eye is known as image steganography. Image steganography can be used in cybersecurity to hide critical information within an image and prevent unauthorized access. Image encryption involves the transformation of an image into a cipher text that cannot be deciphered without a decryption key. Image encryption can be used in the context of cybersecurity to protect private photographs from unauthorized access. The technique of placing a visible or invisible mark on an image to establish ownership or authenticity is known as image watermarking. Image watermarking can be used in cybersecurity to stop unauthorized usage or distribution of images. Image analysis is the method of dissecting images to find out crucial information. Image analysis can be used to find anomalies or suspicious activities in an image in the context of cybersecurity.

Thus, this study suggests securing IoE using advanced Digital Image Processing (DIP) methods and cybersecurity mechanisms. Additionally, this section examines how DIP can recognize and stop cyberattacks on IoE networks and devices. DIP techniques are used to protect the IoE networks; for instance, authors in [16] suggested an approach for finding anomalies in surveillance photos by combining image processing methods with ML algorithms. A framework for object detection and tracking in surveillance films using deep learning techniques is proposed in [17]. Using image processing methods for intelligent transportation systems, authors in [18] suggested an approach for detecting vehicle license plates.

This section presents the IoE security using advanced DIP methods and cybersecurity mechanisms and how DIP can recognize and stop cyberattacks on IoE networks and devices. By analyzing email headers and message content, DIP techniques can spot phishing attempts, detect malware attacks, and spot patterns of harmful activity in network traffic. The suggested architecture can offer a thorough and integrated approach to protecting IoE devices and networks and can act as a guide for upcoming work in this field [19].

DIP approaches can be used to improve cybersecurity in the IoE, but some obstacles must be resolved. The high computational cost of DIP techniques, which might be a bottleneck for real-time applications, is one of the main problems. Additionally, DIP approaches are susceptible to assaults like adversarial attacks, in which a perpetrator alters an image to avoid detection.

Despite these obstacles, DIP approaches also offer many IoE cybersecurity prospects. DIP approaches can be integrated with other cybersecurity measures like encryption and authentication to provide a more reliable security solution. Using DIP approaches can also enable real-time cyberattack detection and reaction, significantly reducing the effect of a security breach.

This study aims to improve the security of IoE devices and networks by applying cutting-edge DIP techniques and cybersecurity measures. The precise objectives are the following:

1. Analyzing email headers and message content can help spot phishing attempts. DIP techniques can also detect malware attacks and identify harmful behavior patterns in network traffic.
2. To determine if cutting-edge cybersecurity methods, such as intrusion detection systems, firewalls, and access control systems, protect IoE devices and networks.
3. To propose a plan for securing IoE devices and networks by combining DIP methods with sophisticated cybersecurity measures.

4. The use of DIP techniques to detect and stop cyberattacks on IoE networks and devices.
5. The efficiency of cutting-edge cybersecurity solutions, such as firewalls, access control systems, and intrusion detection systems, protects IoE devices and networks.
6. The architecture for safeguarding IoE devices and networks combines DIP approaches and cutting-edge cybersecurity solutions.

The study focuses on utilizing and integrating currently available techniques and measures rather than creating new DIP or cybersecurity measures.

6.9 Conclusion

In conclusion, due to the growing number of devices and systems connected to the Internet, protecting the Internet of Everything (IoE) has become a pressing issue. For IoE networks to be protected against cyberattacks and data breaches, advanced digital image processing methods and security structures are crucial. Identifying potential security risks in IoE networks has been proposed using digital image processing techniques, such as object, face, and gesture recognition. To identify unusual behavior or potential security breaches, these algorithms analyze images taken by IoE equipment, such as security cameras. These methods are now more accurate and efficient thanks to the introduction of deep learning algorithms, making them appropriate for usage in practical settings.

The IoE network security also depends on security architectures. These architectures' complete approach to protecting IoE devices and systems includes access control, data encryption, and secure communication protocols. For safeguarding IoE networks, many security designs, including edge computing, fog computing, and blockchain, have been proposed. These architectures make it possible to administer, store, and communicate securely, which increases the security of IoE networks. To stay up with the shifting threat landscape, creating new approaches and structures as IoE networks continue to develop is essential. Modern digital image processing methods and security structures must be used to protect IoE networks from potential security risks. Future research should concentrate on creating more practical and effective methods and structures to improve IoE security.

The IoE network security is a challenging and complex issue, yet cutting-edge digital image processing methods and security designs offer a viable solution to this problem. We can ensure that IoE networks are safe and safeguarded against potential hacker assaults and data breaches by putting these strategies and designs in place.

References

1. Zou Y, Zhu J, Wang X, Hanzo L (2016) A survey on wireless security: Technical challenges, recent advances, and future trends. Proceedings of the IEEE 104 (9):1727–1765. https://doi.org/10.1109/JPROC.2016.2558521
2. ALiero MS, Qureshi KN, Pasha MF, Jeon G (2021) Smart Home Energy Management Systems in Internet of Things networks for green cities demands and services. Environmental Technology & Innovation:101443. https://doi.org/10.1016/j.eti.2021.101443
3. Lakew DS, Dao N-N, Cho S (2022) Adaptive partial offloading and resource harmonization in wireless edge computing-assisted IoE networks. IEEE Transactions on Network Science and Engineering 9 (5):3028–3044. https://doi.org/10.1109/TNSE.2022.3153172
4. Qureshi KN, Alhudhaif A, Haider SW, Majeed S, Jeon G (2022) Secure Data Communication for Wireless Mobile Nodes in Intelligent Transportation Systems. Microprocessors and Microsystems:104501. https://doi.org/10.1016/j.micpro.2022.104501
5. Adhikari M, Munusamy A, Kumar N, Srirama SN (2021) Cybertwin-driven resource provisioning for IoE applications at 6G-enabled edge networks. IEEE Transactions on Industrial Informatics 18 (7):4850–4858. https://doi.org/10.1109/TII.2021.3096672
6. Su J, He S, Wu Y (2022) Features selection and prediction for IoT attacks. High-Confidence Computing 2 (2):100047. https://doi.org/10.1016/j.hcc.2021.100047
7. Qureshi KN, Alhudhaif A, Hussain A, Iqbal S, Jeon G (2021) Trust aware energy management system for smart homes appliances. Computers & Electrical Engineering:107641. https://doi.org/10.1016/j.compeleceng.2021.107641
8. Ullah F, Pun C-M (2022) Enabling Parity Authenticator-Based Public Auditing With Protection of a Valid User Revocation in Cloud. IEEE Transactions on Computational Social Systems. https://doi.org/10.1109/TCSS.2022.3165213
9. Kim J, Caire G, Molisch AF (2015) Quality-aware streaming and scheduling for device-to-device video delivery. IEEE/ACM Transactions on Networking 24 (4):2319–2331. https://doi.org/10.1109/TNET.2015.2452272
10. Thotahewa KMS, Redouté J-M, Yuce MR (2014) Ultra wideband wireless body area networks. Springer,
11. Han W, Wang J, Hou S, Bai T, Jeon G, Rodrigues JJ (2023) An PPG signal and body channel based encryption method for WBANs. Future Generation Computer Systems 141:704–712. https://doi.org/10.1016/j.future.2022.11.020
12. Zhu M, Sui T, Wang R, Sun J (2022) Sensors Scheduling for Remote State Estimation Over an Unslotted CSMA/CA Channel. IEEE Transactions on Network Science and Engineering. https://doi.org/10.1109/TNSE.2022.3210285
13. Sarma J, Biswas R (2023) A power-aware ECG processing node for real-time feature extraction in WBAN. Microprocessors and Microsystems 96:104724. https://doi.org/10.1016/j.micpro.2022.104724
14. Bokhari S, Hamrioui S, Aider M (2022) Cybersecurity strategy under uncertainties for an IoE environment. Journal of Network and Computer Applications 205:103426. https://doi.org/10.1016/j.jnca.2022.103426
15. Magdy M, Hosny KM, Ghali NI, Ghoniemy S (2022) Security of medical images for telemedicine: a systematic review. Multimedia Tools and Applications 81 (18):25101–25145. https://doi.org/10.1007/s11042-022-11956-7
16. Khan SW, Hafeez Q, Khalid MI, Alroobaea R, Hussain S, Iqbal J, Almotiri J, Ullah SS (2022) Anomaly detection in traffic surveillance videos using deep learning. Sensors 22 (17):6563. https://doi.org/10.3390/s22176563

17. Himeur Y, Al-Maadeed S, Kheddar H, Al-Maadeed N, Abualsaud K, Mohamed A, Khattab T (2023) Video surveillance using deep transfer learning and deep domain adaptation: Towards better generalization. Engineering Applications of Artificial Intelligence 119:105698. https://doi.org/10.1016/j.engappai.2022.105698
18. Oliveira-Neto FM, Han LD, Jeong MK (2013) An online self-learning algorithm for license plate matching. IEEE Transactions on intelligent transportation systems 14 (4):1806–1816. https://doi.org/10.1109/TITS.2013.2270107
19. Cai Z, Wan X, Liu X, Ren Q, Lian X, Wang L (2022) Physics-Based Modeling Strategies of Phase-Change Random Access Memory. IEEE Transactions on Electron Devices. https://doi.org/10.1109/TED.2022.3215550

Chapter 7
Machine Learning-Based Detection and Prevention Systems for IoE

Amna Khatoon, Asad Ullah, and Muhammad Yasir

7.1 Overview

The concept of the Internet of Everything (IoE), an interconnected ecosystem of devices, people, data, and processes, has emerged due to the Internet of Things (IoT) rapid proliferation. Machine learning (ML)-based detection and prevention systems can take advantage of the massive amounts of data that this ecosystem produces to boost security, effectiveness, and performance. This chapter investigates the numerous uses and advantages of incorporating ML into the IoE setting. The network of physical objects, including sensors, appliances, and automobiles, connected to the Internet and exchanged data is known as the IoT. The IoE, which broadens the idea to include devices, people, data, and processes, results from the IoT's rapid growth. This interconnected ecosystem's large amounts of data can be used for various purposes, such as ML-based detection and prevention systems. This chapter will examine, through examples, the uses and advantages of incorporating ML into the IoE setting.

A. Khatoon (✉)
Department of Information Engineering, Chang'an University, Xi'an, China
e-mail: 2018024900@chd.edu.cn

A. Ullah
Department of Information Engineering, Chang'an University, Xi'an, China

School of Information Engineering, Eurasia University, Xi'an, China

M. Yasir
Department of Information Engineering, Chang'an University, Xi'an, China

School of Computer and Information Engineering, Henan University, Kaifeng, China

K. Naseer Qureshi et al. (eds.), *Cybersecurity Vigilance and Security Engineering of Internet of Everything*, Internet of Things,
https://doi.org/10.1007/978-3-031-45162-1_7

7.2 Evolution of IoT to IoE

The term IoE refers to the vast network of connected things, people, data, and activities made possible by the Internet [1]. It goes beyond the Internet of Things (IoT), mainly concentrating on tying together and coordinating communication between gadgets and sensors. The IoE aspires to provide a comprehensive ecosystem with interconnected components, enabling better automation, efficiency, and decision-making across many industries. ML-based detection and prevention systems can harness the plethora of data the IoE provides to improve security, effectiveness, and overall performance. By utilizing the interconnection of devices, data, and processes, these systems may identify and resolve possible threats or operational concerns quickly and effectively [2]. The development of the IoT, which consists of countless connected devices talking and exchanging data, has led to the emergence of this idea. The IoE can be used in ML-based detection and prevention systems to improve security, effectiveness, and overall performance.

As the Internet and computer networking technology advanced in the latter half of the twentieth century, the idea of connected devices began to take shape. Researchers and creators started looking at the advantages of connecting actual objects to the Internet. Kevin Ashton first used the phrase "Internet of Things" in a presentation to Procter and Gamble in 1999 [3]. At this point, the extensive debate over the advantages of Internet-connected ordinary objects began. With improvements in wireless communication, sensors, and cloud computing, IoT began to pick up steam around 2000. IoT technology attracted significant investment from businesses like IBM, Cisco, and Google as the number of linked devices increased quickly. As businesses and researchers realized the potential of connecting devices and people, processes, and data starting in 2010, the idea of IoE started to take shape. This extended vision incorporated data analytics, ML, and artificial intelligence (AI) to produce more effective and intelligent systems. Predictive maintenance, smart grids, and personalized marketing are new business models, services, and solutions that the IoE has sparked. As the number of connected devices and the volume of data generated keep expanding, data privacy and security remain significant issues. Governments and regulatory agencies are actively striving to establish standards and laws to solve these issues and encourage the development of the IoE ecosystem. Improvements in connectivity, AI, and ML technologies will enable even more cutting-edge applications and breakthroughs in the IoE in the future. To ensure the ethical and sustainable expansion of the IoE, it will be essential to address the security and ethical issues.

Examining the growth of the IoT and how it influenced the birth of the IoE is crucial to comprehending this evolution. The IoT is a network of connected sensors and gadgets that can communicate and share data. When wireless communication technologies, embedded systems, and the Internet first came together, it made it possible for devices to connect and communicate without human interaction. Through connectivity, various industries, including healthcare, home, and industrial automation, could create intelligent environments, automate tasks, and monitor them remotely [4]. As IoT technologies are more widely used, exponentially more linked devices produced enormous amounts of data. This data opened up new possibilities for extracting insights and streamlining operations across other areas, along with cloud computing and data analytics developments. At this point, connecting devices and utilizing the data are the key areas of concentration. As the IoT grew, it is clear that a broader perspective was required to tap into the interconnected ecosystem's potential effectively. As a result, the IoE was born, expanding the idea beyond only devices and sensors to encompass people, data, and processes. The IoE values how people engage with technology and works to integrate all facets of the ecosystem to build settings that are more effective, intelligent, and responsive.

Moving from the IoT to the IoE requires not only the blending of systems, people, data, and processes but also the application of sophisticated analytics and artificial intelligence (AI) methods like ML. ML algorithms can find patterns, trends, and insights by sifting through the massive amounts of data produced by the IoE, leading to better user experiences, more effective decision-making, and increased efficiency [5]. The transition from IoT to IoE has made it possible to implement various new technologies in several different industries, including smart cities, healthcare, agriculture, transportation, and energy management. With the increased flexibility, scalability, and adaptability provided by the IoE, it is now feasible to tackle complicated problems and build more sustainable, effective, and connected environments.

Intrusion detection and prevention systems (IDPS) aim to detect and stop unauthorized access, security breaches, and other harmful activities. These systems monitor network traffic, system activity, and user behavior [6]. The security and integrity of computer networks, programs, and data depend on these systems. ML algorithms can improve the efficacy of intrusion detection and prevention systems by automating the processing of massive amounts of data and spotting trends related to cyber threats. IDPS often fall into one of two categories including network-based IDPS and host-based IDPS systems.

Network-based (NIDPS) systems are used to monitor the network traffic to spot and stop harmful actions like intrusion attempts, malware infections, and distributed denial-of-service (DDoS) attacks. These systems examine the network packets for known attack signatures or potentially destructive behavioral patterns. On the other hand, the host-based HIDPS systems concentrate on specific hardware or hosts, monitoring user activity, application activity, and system logs to identify and stop unauthorized access, malware infections, or other security breaches. To spot indications of compromise, HIDPS can also keep track of alterations to essential system files, registry settings, or other configuration information.

7.3 Importance of Machine Learning in IoE

The IoE ecosystem depends on ML because it gives us the tools to analyze the massive amounts of data produced by linked things, people, and processes [7]. By incorporating ML into the IoE, numerous industries can make better decisions and operate more efficiently. These networked devices capture much data, which ML algorithms may analyze to spot patterns, trends, and abnormalities. ML has the potential to advance the following areas in detection and preventive systems:

- ML algorithms can analyze information from various devices and sensors to spot odd or suspicious trends. The system may immediately identify potential security threats or operational problems and take appropriate action by detecting anomalies in real time.
- ML models can be trained to recognize patterns of activity that point to unauthorized access or intrusion attempts for intrusion detection and prevention. After that, these models may monitor network activity and device behavior, alerting users or denying them access when they notice questionable behavior.
- Using data gathered from sensors and devices, ML can be used to forecast equipment breakdowns and plan maintenance. It assists in lowering downtime and enhancing overall operational effectiveness.
- By detecting and preventing unwanted access and locating potential weaknesses inside the linked devices, ML algorithms can assist in protecting sensitive data.
- ML models can be used to anonymize data or apply privacy-preserving measures, ensuring that sensitive or private information is kept secure while enabling analysis and decision-making with the data.
- By examining past data, ML can assist organizations in creating more effective incident response plans, allowing them to react to security problems or operational disruptions more quickly.
- ML models can spot patterns of fraudulent conduct, including odd financial transactions, aiding enterprises in real-time fraud detection and prevention.

The development of intelligent systems has been profoundly impacted by the combination of ML and the IoE, allowing for more effective automation and decision-making. Early connected devices produced little data, and the IoE used ML sparingly because there was not enough data for training and analysis. Early in the IoT, descriptive statistics and reporting dominated data analysis, with little to no usage of ML approaches. Because they lacked ML algorithms' intelligence, early IoE systems were more rigid and less adaptive. The IoE ecosystem currently produces enormous amounts of data from numerous sources, offering ample ML application opportunities. With the ability to collect, analyze, and understand vast volumes of data, ML has emerged as a crucial element of the IoE [8]. This allows for better automation and decision-making. Predictive maintenance, anomaly detection, personalized marketing, and traffic management are just a few IoE applications that use ML. With quicker data processing at the source made possible by the shift to edge computing, ML-powered IoE systems are more effective and responsive.

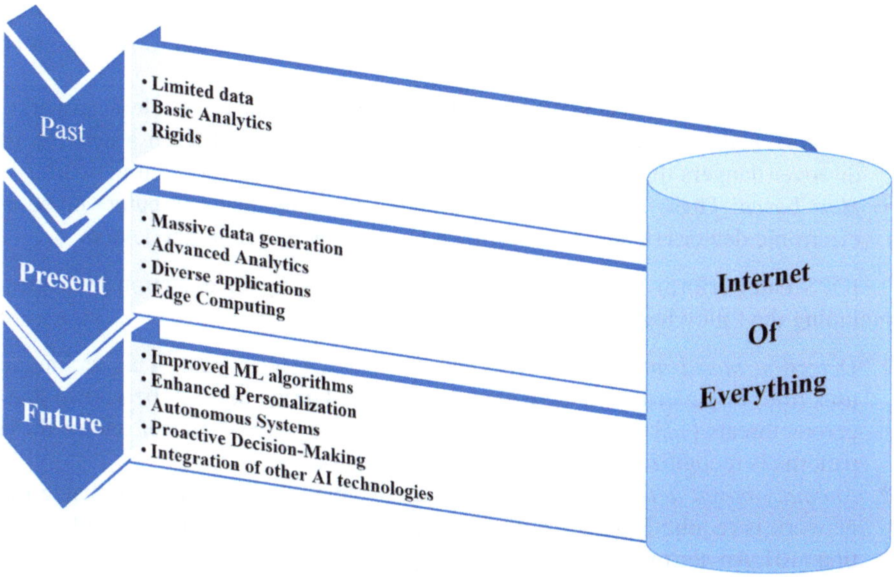

Fig. 7.1 ML and IoE

Combining ML and IoE will provide even more innovative and effective systems as ML algorithms advance. Users will enjoy more individualized interactions due to ML approaches because the IoE system will respond differently to each user's tastes and needs. More autonomous systems, e.g., self-driving cars and drones, that can make judgments and take action without human intervention will emerge due to the merging of ML with IoE. Systems can make proactive decisions thanks to ML and IoE, which will increase efficiency and provide better user experiences by foreseeing and anticipating user wants and system requirements. Future IoE applications will be more sophisticated and advanced due to combining IoE with other AI technologies like computer vision and natural language processing (NLP). From early iterations of sparse data and crude analytics to the sophisticated applications and intelligent systems of the present, the combination of ML and the IoE has seen significant development [9, 10]. As AI technology advances and the IoE ecosystem expands, the future of ML and IoE integration promises to provide increasingly more sophisticated and autonomous applications, as in Fig. 7.1.

7.4 Intrusion Detection and Prevention Systems

A security system known as an IDPS monitors and analyzes the traffic on a network in search of indications of hostile behavior and potential dangers. It is designed to detect and prevent cyberattacks by identifying suspicious behavior and stopping illegal actions [11]. There are four basic categories of IDPS, which are as follows:

- *Network-based*: These are installed at critical locations within the network to monitor incoming and outgoing traffic to all devices. They are also known as network-based firewalls.
- *Wireless*: These are created expressly to monitor the traffic on wireless networks.
- *Network behavior analysis*: These systems analyze the traffic on a network to discover dangers that cause traffic flows that are not typical for that network.
- *Host-based*: These applications are installed on individual computers or other electronic devices (hosts) to monitor and analyze activity on that device.

IDPS can perform its functions by utilizing a variety of detection strategies, including the following:

1. *Detection based on predefined rules or patterns (signatures)*: This technique uses rules or patterns that have already been predefined to describe known dangerous threats [12]. An alert is generated by the system whenever one of these patterns is recognized in the traffic on the network.
2. *Anomaly-based detection*: Establishing a baseline of typical activity inside the network is required for this detection method, known as anomaly-based detection [13]. An alarm is generated by the system whenever the observed behavior significantly deviates from this baseline in some way.
3. *Policy-based detection*: This method employs distinct security policies governed by the administrator's configuration settings. The security system will alarm if it detects any actions against these policies.
4. *Hybrid detection*: This technique combines signature-based and anomaly-based detection to provide a more all-encompassing detection of potential threats.

IDPS solutions increasingly use ML to improve their threat detection capabilities. This trend is expected to continue. Traditional IDPS systems call for manual updates to their signature databases. They frequently struggle to recognize zero-day vulnerabilities (new threats not seen before) and complicated attacks that include many stages [14]. The IDPS can "learn" from the data it examines, made possible through ML, which helps address these constraints. Several applications of machine learning are possible within the framework of IDPS, including the following:

1. *Anomaly detection*: ML algorithms can be trained to understand what "normal" activity looks like in a network. It enables anomaly detection. They are then able to recognize and report unusual activity to administrators, which may signal the presence of a threat.
2. *Predictive analysis*: ML can assist in predicting future assaults based on previous data and emerging trends. This type of analysis is used in predictive analysis.
3. *Reducing the number of false-positive warnings*: Machine learning can help differentiate between actual threats and harmless abnormalities, which can assist in reducing the number of false-positive warnings.
4. *Automated response*: ML can help design automated responses to specific kinds of *threats*, enhancing the "prevention" component of an IDPS.

However, despite ML improving IDPS capabilities, it is essential to remember that ML solutions still require human oversight. Because they can produce both false positives and false negatives, human control is necessary to ensure that the system is operating as effectively as possible.

7.5 Existing IDPS Solutions Designed for IoE Networks

Several existing IDPS solutions are designed specifically for IoE networks.

Cisco Stealthwatch Cisco Secure Network Analytics, a comprehensive network traffic analysis and detection solution, employs advanced ML algorithms and behavioral analysis through its IDPS Stealthwatch [15]. It extends real-time threat detection to the entire network, including IoE devices. Its unique, multilayered approach enables sophisticated threat detection, rapid response, and more straightforward network segmentation, ensuring continuous awareness of network activities. Through its agentless design, visibility is extended from on-premises to the cloud, with the capability to identify malware in encrypted traffic and ensure policy compliance without decryption. The integration with the Cisco SecureX platform expands the benefits of secure network analytics beyond the network and cloud to endpoints and applications, providing a holistic security solution for the digital enterprise.

Palo Alto Networks IoT Security Palo Alto Networks provides IoT security solutions that excel in advanced threat detection and prevention for IoE networks. These solutions are adept at identifying abnormal behavior and safeguarding IoT devices from many attacks, harnessing the power of ML-based analytics. The cornerstone of Palo Alto's IoT Security solution is its seamless integration with next-generation firewalls. This symbiosis enables the dynamic discovery and real-time maintenance of an inventory of the IoT devices on your network. This ongoing and automated inventory management is an indispensable feature, ensuring that your IoT device landscape is always accurately represented and up-to-date. Their innovative AI and ML algorithms set Palo Alto's IoT Security solutions apart. These sophisticated technologies confer the solution with exceptional accuracy [16]. They even enable the precise classification of previously unencountered IoT device types, enhancing the system's adaptability and preparedness for emerging threats. In addition to its core functionalities, IoT Security offers added benefits such as the automatic generation of policy recommendations. These recommendations are instrumental in controlling IoT device traffic, thereby minimizing the risk of breaches and ensuring smoother network operations. Moreover, the solution facilitates the automatic creation of IoT device attributes, which can be incorporated into firewall policies for a more robust and tailored security approach. An IoT Security subscription is required to access all these features and fortify your IoE networks. This investment ensures that your network is equipped with the most advanced tools and technologies to secure your IoT ecosystem against existing and emerging threats.

Darktrace Industrial Immune System Darktrace, a leading cyber-threat defense company, has developed a revolutionary product, the Industrial Immune System, to bolster the protection of critical infrastructure, including Industrial Control Systems (ICS) and SCADA (Supervisory Control and Data Acquisition) systems. These systems, which manage critical processes in sectors like power generation and manufacturing, are increasingly targeted by sophisticated hackers, posing a growing security challenge. Using ML algorithms developed at the University of Cambridge, the Industrial Immune System detects emerging cyber-threats and abnormal behavior within these critical environments in real time. This product offers a holistic, visual overview of production environments, enabling security operators to anticipate and mitigate potential threats before they escalate into severe cyberattacks. This unique approach forms an adaptive "pattern of life" for machines, networks, and users within these environments, significantly improving the resilience of critical infrastructure [8]. The Industrial Immune System is part of Darktrace's flagship Enterprise Immune System and has been successfully implemented by major players, such as European energy leader Drax. The system continuously monitors and alerts to suspicious or abnormal activity within corporate IT and SCADA networks. This technology has significantly transformed threat detection and prevention capabilities, enhancing operational security and resilience [17]. To learn more about Darktrace's Industrial Immune System, visit their website, or contact them directly via email. Darktrace, founded in 2013 by ML specialists and government intelligence experts, is recognized as one of the world's leading cyber-threat defense companies. They have a global presence with headquarters in Cambridge, UK, and Washington DC and offices worldwide.

Trend Micro Tipping-Point Trend Micro's TippingPoint, an IDPS solution, employs ML and threat intelligence to deliver robust protection for IoE networks, including IoT devices. TippingPoint's capabilities extend beyond just detecting known threats. It is adept at identifying and blocking even unknown attacks, thus providing an added layer of security to your network. One of the cornerstone features of TippingPoint is its ability to provide complete visibility across your network. Armed with insightful and contextual data, it can measure and drive vulnerability threat prioritization, ensuring that the most severe threats are addressed promptly [2]. Deep network traffic inspection helps identify and block threats often undetected by traditional security solutions. Additionally, TippingPoint incorporates an on-box SSL inspection feature. It reduces the security created by encrypted traffic, further bolstering the security of your network. TippingPoint offers flexible deployment options that are easy to set up and manage. With a centralized management interface, it simplifies security operations, making it a user-friendly solution even for those without extensive technical expertise. The solution provides immediate and ongoing threat protection with out-of-the-box recommended settings, ensuring your network remains secure. TippingPoint also introduces several features to improve operational efficiency. These include reassigning licenses across TippingPoint System (TPS) deployments without changing network infrastructure and a pay-as-you-grow licensing model. The latter allows you to scale performance and security requirements quickly as your network grows and evolves. These features make TippingPoint a robust, flexible, cost-effective security solution. In summary, Trend Micro's TippingPoint is a comprehensive IDPS solution that combines advanced threat detection and mitigation capabilities with operational efficiency. By offering complete network visibility, deep traffic inspection, and flexible deployment options, it provides a robust and scalable security solution for IoE networks.

Fortinet FortiSIEM Fortinet's FortiSIEM is a comprehensive Security Information and Event Management (SIEM) solution with ML-based analytics. This tool is a backbone to security operations teams, providing the necessary capabilities for in-depth threat detection and response within IoE networks [18]. The core of this solution is a fully integrated configuration management database (CMDB), unique within the industry. The CMDB generates a complete asset inventory through active and passive discovery methods, tracking the states of devices and applications over time. It consistently collects critical information, providing real-time knowledge of the operational environment and enabling teams to address issues proactively [19]. The inclusion of FortiGuard Labs, a 24/7 threat analysis and mitigation operation, enhances the threat detection capabilities of FortiSIEM. This solution defends against known threats and employs AI-driven anomaly detection capabilities, such as User and Entity Behavior Analytics (UEBA), to counteract unknown threats [20]. A new feature that fortifies the solution is visual threat hunting through link analysis, which enables clear visualization of relationships between users, devices, and incidents. FortiSIEM provides various features and benefits, from self-learning asset inventory and real-time security analytics to streamlined investigations and continuous compliance. It seamlessly integrates with Fortinet's portfolio and third-party solutions, delivering deep fabric integration [21]. FortiSIEM is adaptable and serves various use cases such as converged IT/OT SOC, remote operations, on-premises deployment, and multi-cloud environments. It also supports a hybrid approach, combining software as a Service (SaaS), cloud, virtual machines [22], and hardware to suit any needs. In conclusion, Fortinet's FortiSIEM offers an advanced and robust SIEM solution, ensuring a high level of security for today's complex IoE networks.

Symantec IoT Security Symantec provides IoT Security solutions, utilizing ML to protect IoE networks from various threats. One such offering is the ICSP Neural USB scanning station, an appliance designed to detect and prevent USB-borne malware in IoT environments. ICSP Neural incorporates deep learning capabilities, ensuring future-proof protection against evolving attacks. ICSP Neural's lightweight enforcement driver, compatible with legacy systems, permits only scanned external media on your systems, enhancing overall security. The appliance is also widely compatible with automation vendors, healthcare devices, and defense systems, ensuring robust protection against known and unknown threats. Symantec offers Critical System Protection (CSP) for legacy or end-of-life systems, which provides offline-capable intrusion prevention and detection features for IoT devices [23]. Using a signature-less, policy-based approach, CSP ensures endpoint security and compliance, protecting IoT devices from exploits and attacks. Symantec's solutions address industry challenges by offering robust security without replacing existing infrastructure. The combination of ICSP Neural and CSP provides comprehensive protection against USB-borne malware, network intrusions, and zero-day exploits targeting industrial control systems and IoT devices.

FireEye Helix FireEye Helix is a cutting-edge, cloud-based security platform that employs ML and advanced analytics to detect and respond to threats within the IoE networks. The platform presents a holistic, foundational approach to cybersecurity, offering real-time visibility, intelligent threat detection, and automated incident response capabilities. It helps organizations establish a robust security foundation amid the escalating complexity of modern cyber-threats. The sophistication of cybersecurity threats today can expose companies' vulnerabilities almost daily, prompting them to invest more in products and talent. However, this often reactive approach only compounds the complexity, presenting another opportunity for attackers. FireEye Helix is designed to circumvent this issue by making it simple to provide advanced security to any organization [24]. Helix leverages frontline intelligence to identify unseen threats, informing expert decision-making and enabling organizations to utilize their security investments fully. It collates event data from FireEye and non-FireEye components of an organization's security infrastructure. This data is combined with frontline intelligence, rules, and analytics to provide the context necessary to determine the most severe threats and inform response strategies. FireEye Helix streamlines all Security Operation Center (SOC) functions with a single interface. It includes alert management, search, analysis, investigations, and reporting, which empowers organizations to regain control over their cybersecurity measures and minimize potential risks.

Check Point IoT Security Check Point's Quantum IoT Protect is a robust IoT Security solution designed explicitly for IoE networks. It incorporates unified threat management and advanced threat prevention, leveraging ML algorithms to identify and counter IoT-specific threats. The goal is to protect connected devices across various environments, such as enterprise smart offices, smart buildings, industrial, and healthcare. The rise in IoT device utilization across enterprises, industrial organizations, and healthcare sectors underscores the urgency of a comprehensive security solution. Statistics show that 63% of enterprises, 92% of industrial organizations, and 82% of healthcare organizations utilize IoT [25]. While IoT devices offer significant operational benefits, their connection to the network simultaneously extends the attack surface, providing more entry points for potential cyberattacks. Quantum IoT Protect is tailored to minimize your organization's exposure to IoT cyber-risks and effectively thwart cyberattacks in response to this escalating risk. If your organization develops or deploys IoT devices, our solutions are designed to offer maximum security. To pre-empt IoT cyberattacks, such as phishing, ransomware, and crypto mining, Quantum IoT Protect offers strategic approaches, including identifying IoT security risks across IT/OT networks, developing rapid policies to secure IoT devices, and preventing network-based and device-level attacks. These strategies are essential to maintaining a secure IoT network and preventing potential security breaches.

IBM Q-Radar IBM Q-Radar is a Security Information and Event Management (SIEM) solution that uses ML analytics to detect and respond to threats in the IoE networks. It offers real-time visibility, anomaly detection, and automated incident response capabilities, thus creating a robust defense against potential threats. The IBM Security Q-Radar Suite is a state-of-the-art threat detection and response solution engineered to streamline the experience of security analysts and hasten their pace across the complete incident lifecycle. With enterprise-grade artificial intelligence (AI) and automation embedded, the suite significantly enhances analyst productivity, aiding security teams that are constrained in resources to work more effectively across core technologies. This suite provides integrated products for endpoint security, including Endpoint Detection and Response (EDR), Extended Detection and Response (XDR), and Managed Detection and Response (MDR), as well as log management, SIEM and Security Orchestration, Automation, and Response (SOAR). These elements share a familiar user interface, insights, and connected workflows. An intuitive user interface enables analysts to work swiftly and efficiently through their investigation and response processes. Shared insights and automated actions across products further enhance this efficiency. The suite's unique enterprise-grade AI capabilities allow analysts to automatically contextualize and prioritize threats, increasing the speed and accuracy of their response. Delivered as a service on AWS, IBM Security QRadar Suite products facilitate simple deployment across cloud environments and integration with public cloud and software as a Service (SaaS) log data [26]. Thanks to its new cloud-native security observability and log management capability, the suite is optimized for large-scale data ingestion, subsecond search, and rapid analytics. IBM Security QRadar Suite is designed as an open platform with over 900 pre-built integrations for flexibility across IBM and third-party products. This open platform approach combines core technologies required in today's Security Operation Centers (SOCs). It includes native, pre-integrated capabilities for Threat Intelligence, Log Management, EDR, SIEM, and SOAR, providing a comprehensive, integrated approach to network security.

Juniper Networks Sky Advanced Threat Prevention Juniper Networks' Sky Advanced Threat Prevention (ATP) Cloud is a sophisticated solution that safeguards IoE networks from advanced threats. Leveraging ML algorithms and threat intelligence, it proficiently detects and blocks malicious activities aimed at IoT devices [27]. ATP Cloud can be used as an additional SRX Series Firewall license. This unique integration facilitates a blend of static and dynamic analysis with ML, facilitating rapid identification of unknown threats downloaded from the web or transmitted via email. Following the assessment, ATP Cloud sends a file verdict and risk score back to the SRX Series firewall, enabling efficient blocking at the network level. Further, ATP Cloud supplies SecIntel security intelligence feeds encompassing malicious domains, URLs, and IP addresses. These feeds are derived from a comprehensive file analysis, research conducted by Juniper Threat Labs, and trusted third-party threat feeds. SRX Series Firewalls and Juniper Networks MX Series Universal Routing Platforms automatically collect and distribute these feeds to obstruct command-and-control communications, enhancing the organization's security posture. Crucial insights into DNS traffic on your network are another advantage of ATP Cloud. It provides critical information to counteract attacks that utilize DNS for command and control or to transfer and exfiltrate data. ATP Cloud offers protection against threats from DNS-generating algorithms (DGA) and DNS tunneling. To tackle security issues arising from the widespread use of IoT, ATP Cloud can identify and classify IoT devices on the network. This classification allows security operations teams to manage feeds for policy enforcement across the network and mitigate the risks of extensive IoT attack surfaces. When considering IDPS solutions for IoE networks, assessing your network's specific requirements and characteristics is paramount. This evaluation will ensure you select the most fitting solution for your organization's needs.

7.6 Attacks in IoE Networks and Its Pretension by Using ML-Based IDPS Systems

The Internet of Things (IoT) and the IoE networks provide substantial problems to network security due to the enormous number of networked devices and the various communication protocols they use. As a result of this circumstance, these networks are open to a wide variety of cyberattacks. On the other hand, IDPS powered by ML can be used to identify and stop these kinds of dangers.

- *Botnets*: Botnets are networks of Internet of Things devices that have been compromised and are controlled by an adversarial actor. An ML-based IDPS is responsible for monitoring the traffic patterns of a network and identifying any anomalous activity that may indicate the presence of a botnet [22]. Using ML methods, these systems can learn from previously collected data to identify new or emerging botnet varieties.
- *Denial-of-service (DoS) attacks*: A DoS attack aims to overwhelm the resources of a network or a specific device, preventing that device from functioning normally. ML-based intrusion detection and prevention systems analyze traffic patterns and look for abrupt spikes or anomalies in network traffic, which could indicate a DoS assault. These systems, which use algorithms for ML, can differentiate between legal and malicious communications, which enables immediate corrective steps to be taken.
- *Malware propagation*: Malicious software that targets IoT devices can quickly move over IoE networks, putting the linked devices' privacy and security at risk. ML-based IDPS conduct behavior analysis on connected devices to identify odd behaviors that may indicate the presence of malware [15]. These ML models,

trained on known malware behaviors, can recognize and stop efforts at malware spread.

- *Impersonation of devices and spoofing*: Attackers may try to gain unauthorized access to IoE networks by pretending to be genuine via spoofing or impersonating those devices. ML-based IDPS learn the behavior patterns of connected devices and look for anomalies that deviate from typical device attributes. It allows the IDPS to recognize odd behavior, which may indicate that the device is being spoofed.
- *Attacks from the inside*: An inside attack on an IoE network could be the result of acts taken on purpose by authorized users or devices that have been compromised. Establishing baselines of usual user behavior and identifying deviations from these patterns can be done with ML-based IDPS. These systems can identify potential insider threats by monitoring user actions and applying ML algorithms. They can then execute appropriate corrective steps. When defending against IoE network assaults, ML-based IDPS systems have several distinct advantages. They can adjust and improve by continually gaining knowledge from new attack patterns and constantly changing dangers. However, it is necessary to realize that ML-based IDPS systems are imperfect and may have flaws, such as false positives or negatives. It is one of the drawbacks that must be acknowledged [28]. As a result, a comprehensive security plan for IoE networks should incorporate ML-based solutions in addition to other security measures such as tight authentication procedures and network segmentation.

7.7 Pros and Cons of ML in Detection and Prevention Systems

Cybersecurity, environmental monitoring, fraud detection, and supply chain management are just a few industries that can transform by integrating the IoE and ML in detection and prevention systems. However, this integration has advantages and disadvantages of its own. For ML-based detection and prevention systems, the following is a summary of the benefits and drawbacks of IoE:

Pros

1. ML algorithms can analyze vast amounts of data from networks, linked devices, and sensors, producing predictions and insights that are more precise. It allows businesses to streamline operations, cut waste, and boost productivity.
2. The IoE makes gathering and analyzing data in real time more accessible, enabling businesses to identify possible dangers, abnormalities, and problems earlier and take quick action. It ensures the efficient running of systems and processes while reducing damage and downtime.

3. ML-based systems can forecast possible problems or trends, allowing businesses to be proactive and take preventive action. It can lower expenses, lessen hazards, and boost overall performance.
4. The IoE makes integrating new devices and sensors into the network simple, enabling businesses to grow their ML-based systems to meet shifting demands.
5. As more data becomes accessible, ML systems can learn and adapt over time, increasing their efficiency and precision. It enables businesses to stay on top of the quickly evolving environment and potential dangers or difficulties.

Cons

1. The IoE includes gathering, storing, and analyzing enormous volumes of data, which raises questions regarding user privacy and data security. Organizations using IoE-based ML systems have significant hurdles in protecting sensitive data and managing privacy issues.
2. Integrating ML algorithms with IoE technologies can be difficult, necessitating businesses to spend money on specialized resources and personnel to manage and maintain these systems.
3. Developing ML algorithms and models and purchasing connected devices, sensors, and infrastructure might come at a high initial cost when adopting and deploying IoE-based ML systems.
4. The dependability of the networks and associated devices affects the performance of IoE-based ML systems. These systems' effectiveness and efficiency may suffer from connectivity problems, device malfunctions, or network outages.
5. Concerns about bias in ML models or unintended outcomes of automated decision-making are some ethical and legal issues that ML algorithms and IoE technologies bring up. When deploying IoE-based ML systems, organizations must navigate these issues.

In conclusion, incorporating IoE and ML into detection and prevention systems has numerous benefits, including increased effectiveness, real-time analysis, proactive decision-making, scalability, and continuous learning. However, it has drawbacks, including privacy and data security issues, complexity, expensive implementation, connectivity problems, reliability issues, and ethical and legal issues. To ensure these technologies' successful deployment and operation, organizations must carefully consider the benefits and drawbacks of using IoE-based ML systems and take steps to resolve these issues.

7.8 Conclusion

Utilizing the IoE for ML-IDPS requires developing and deploying scalable, secure, and efficient data processing and transmission infrastructures. Encourage collaboration across various stakeholders, including governmental organizations, commercial partners, and research institutions, to stimulate innovation and overcome the challenges of implementing and maintaining IoE systems. IoE and ML integration in detection and prevention systems can revolutionize several industries and sectors, including cybersecurity, environmental monitoring, fraud detection, supply chain management, and healthcare. Organizations can increase productivity, increase accuracy, enable real-time analysis, and support proactive decision-making by utilizing the power of ML algorithms and the connectivity of the IoE.

Adopting IoE-based ML systems has several difficulties, including issues with connectivity and stability, data security and privacy, complexity, high implementation costs, and moral and legal dilemmas. To guarantee the effective deployment and operation of IoE-based ML systems, organizations must carefully weigh the advantages and disadvantages, invest in the appropriate knowledge and infrastructure, and address these issues. The potential for ML-based detection and prevention systems in the IoE ecosystem should increase as technologies develop. More complex applications and solutions will be made possible by advancements in AI, communication technologies, and the IoE. Ultimately, it will spur efficiency, sustainability, and safety advancements across numerous industries, revolutionizing how businesses run and deal with problems in a connected world.

References

1. Miraz MH, Ali M, Excell PS, Picking R (2015) A review on Internet of Things (IoT), Internet of everything (IoE) and Internet of nano things (IoNT). 2015 Internet Technologies and Applications (ITA): 219–224. https://doi.org/10.1109/ITechA.2015.7317398
2. Ahmad Z, Shahid Khan A, Wai Shiang C, Abdullah J, Ahmad F (2021) Network intrusion detection system: A systematic study of machine learning and deep learning approaches. Transactions on Emerging Telecommunications Technologies 32 (1):e4150. https://doi.org/10.1002/ett.4150
3. Liu H, Lang B (2019) Machine learning and deep learning methods for intrusion detection systems: A survey. Applied Sciences 9 (20):4396. https://doi.org/10.3390/app9204396
4. Kone R, Toutsop O, Thierry KW, Kornegay K, Falaye J Adversarial Machine Learning Attacks in Internet of Things Systems. In: 2022 IEEE Applied Imagery Pattern Recognition Workshop (AIPR), 2022. IEEE, pp 1–7. https://doi.org/10.1109/AIPR57179.2022.10092216
5. Liu Z, Hao J (2019) Intention recognition in physical human-robot interaction based on radial basis function neural network. Journal of Robotics 2019:1–8. https://doi.org/10.1155/2019/4141269
6. Escobedo C, Strong M, West M, Aramburu A, Roncone A Contact anticipation for physical human–robot interaction with robotic manipulators using onboard proximity sensors. In: 2021 IEEE/RSJ International Conference on Intelligent Robots and Systems (IROS), 2021. IEEE, pp 7255–7262. https://doi.org/10.1109/IROS51168.2021.9636130
7. Al-Haija QA, Alsulami AA (2021) High performance classification model to identify ransomware payments for heterogeneous Bitcoin networks. Electronics 10 (17):2113. https://doi.org/10.3390/electronics10172113
8. Farias da Costa VC, Oliveira L, de Souza J (2021) Internet of everything (IoE) taxonomies: A survey and a novel knowledge-based taxonomy. Sensors 21 (2):568. https://doi.org/10.3390/s21020568
9. Shah J, Malik DL (2013) Impact of DDOS attacks on cloud environment. International Journal of Research in Computer and Communication Technology 2 (7)

10. Haddadin S, De Luca A, Albu-Schäffer A (2017) Robot collisions: A survey on detection, isolation, and identification. IEEE Transactions on Robotics 33 (6):1292–1312. https://doi.org/10.1109/TRO.2017.2723903
11. Qureshi KN, Rana SS, Ahmed A, Jeon G (2020) A novel and secure attacks detection framework for smart cities industrial internet of things. Sustainable Cities and Society 61:102343
12. Robla-Gómez S, Becerra VM, Llata JR, Gonzalez-Sarabia E, Torre-Ferrero C, Perez-Oria J (2017) Working together: A review on safe human-robot collaboration in industrial environments. IEEE Access 5:26754–26773. https://doi.org/10.1109/ACCESS.2017.2773127
13. Zaib MH, Bashir F, Qureshi KN, Kausar S, Rizwan M, Jeon G (2021) Deep learning based cyber bullying early detection using distributed denial of service flow. Multimedia Systems: 1–20
14. AlMasri T, Snober MA, Al-Haija QA IDPS-SDN-ML: An Intrusion Detection and Prevention System Using Software-Defined Networks and Machine Learning. In: 2022 1st International Conference on Smart Technology, Applied Informatics, and Engineering (APICS), 2022. IEEE, pp 133–137. https://doi.org/10.1109/APICS56469.2022.9918804.
15. Kanafi FS, Arnarson H, Bremdal BA A new inexpensive approach for securing cyber-physical systems. In: 2022 IEEE/SICE International Symposium on System Integration (SII), 2022. IEEE, pp 790–796. https://doi.org/10.1109/SII52469.2022.9708861
16. Masoud M, Jaradat Y, Manasrah A, Jannoud I (2019) Sensors of smart devices in the internet of everything (IoE) era: big opportunities and massive doubts. Journal of Sensors 2019. https://doi.org/10.1155/2019/6514520
17. Othman S, Ba-Alwi F, Alsohybe N, Al-Hashida A Intrusion detection model using machine learning algorithm on Big Data environment. J. Big Data 5 (1), 1–12 (2018). https://doi.org/10.1186/s40537-018-0145-4
18. Yang L, Shami A (2022) IDS-ML: An open source code for Intrusion Detection System development using Machine Learning. Software Impacts 14:100446. https://doi.org/10.1016/j.simpa.2022.100446
19. Sianaki OA, Yousefi A, Tabesh AR, Mahdavi M Internet of everything and machine learning applications: issues and challenges. In: 2018 32nd International Conference on Advanced Information Networking and Applications Workshops (WAINA), 2018. IEEE, pp 704–708. https://doi.org/10.1109/WAINA.2018.00171
20. Illy P, Kaddoum G, Kaur K, Garg S (2022) ML-based IDPS enhancement with complementary features for home IoT networks. IEEE Transactions on Network and Service Management 19 (2):772–783. https://doi.org/10.1109/TNSM.2022.3141942.
21. Birkinshaw C, Rouka E, Vassilakis VG (2019) Implementing an intrusion detection and prevention system using software-defined networking: Defending against port-scanning and denial-of-service attacks. Journal of Network and Computer Applications 136:71–85. https://doi.org/10.1016/j.jnca.2019.03.005
22. Ali M, Qureshi KN, Newe T, Aman K, Ibrahim AO, Almujaly M, Nagmeldin W (2023) Decision-Based Routing for Unmanned Aerial Vehicles and Internet of Things Networks. Applied Sciences 13 (4):2131
23. Arowolo MO, Ogundokun RO, Misra S, Agboola BD, Gupta B (2023) Machine learning-based IoT system for COVID-19 epidemics. Computing 105 (4):831–847. https://doi.org/10.1007/s00607-022-01057-6
24. Al-Ayyoub M, Jararweh Y, Benkhelifa E, Vouk M, Rindos A Sdsecurity: A software defined security experimental framework. In: 2015 IEEE international conference on communication workshop (ICCW), 2015. IEEE, pp 1871–1876. https://doi.org/10.1109/ICCW.2015.7247453
25. Wylde V, Rawindaran N, Lawrence J, Balasubramanian R, Prakash E, Jayal A, Khan I, Hewage C, Platts J (2022) Cybersecurity, data privacy and blockchain: A review. SN Computer Science 3 (2):127. https://doi.org/10.1007/s42979-022-01020-4
26. Kone R, Toutsop O, Thierry KW, Kornegay K, Falaye J Adversarial Machine Learning Attacks in Internet of Things Systems. In: 2022 IEEE Applied Imagery Pattern Recognition Workshop (AIPR), 2022. IEEE, pp 1–7. https://doi.org/10.1109/AIPR57179.2022.10092216

27. Sahani N, Zhu R, Cho J-H, Liu C-C (2023) Machine Learning-based Intrusion Detection for Smart Grid Computing: A Survey. ACM Transactions on Cyber-Physical Systems. https://doi.org/10.1145/3578366
28. Haque MS, Ramesh M, Oliveira J P; & Gomide, ADA (2021). Building administrative capacity for development: limits and prospects. International Review of Administrative Sciences 87 (2):211–219. https://doi.org/10.1177/00208523211002605

Chapter 8
Role of Blockchain for IoE Infrastructures and Applications

Ibrahim Tariq Javed and Kashif Naseer Qureshi

8.1 Internet of Everything

The Internet of Everything (IoE) can alter industries, enhance people's daily lives, and provide new opportunities for innovation and growth by utilizing the power of the Internet and the enormous volumes of data generated by this ecosystem. All tangible items and gadgets are linked to the Internet including smartphones, wearables, automobiles, appliances, sensors, and actuators. These devices capture and transmit data to enable smooth communication and engagement within the ecosystem. The IoE puts people at the heart of the ecosystem, emphasizing their interactions with tools, information, and procedures. It covers how individuals communicate, work together, and make decisions using technology. It also considers the ecosystem's social aspects, such as information exchange and community development. Data is crucial to the IoE since it underpins rational decision-making and powers the ecosystem's intelligence. Devices can provide data, other sources can gather it, or humans can create it by interacting with the ecosystem. Identifying patterns, trends, and insights from this data makes improving decision-making and process efficiency possible. People played a primarily passive user role in the IoE ecosystem. The IoE ecosystem now includes individuals as both users and decision-makers. They use voice commands, gestures, and touchscreens to engage more intricately with connected devices and applications. Individuals can contribute data to the

I. T. Javed
Blockchain@UBC, University of British Columbia, Vancouver, BC, Canada
e-mail: IbrahimTariq.Javed@ubc.ca

K. N. Qureshi (✉)
Department of Electronic & Computer Engineering, University of Limerick (UL), Limerick, Ireland
e-mail: kashifnaseer.qureshi@ul.ie

K. Naseer Qureshi et al. (eds.), *Cybersecurity Vigilance and Security Engineering of Internet of Everything*, Internet of Things,
https://doi.org/10.1007/978-3-031-45162-1_8

system through wearables and other personal devices, enabling more individualized experiences and better decision-making.

Early connected device operations are straightforward and restricted to the most fundamental data transmission and control operations. Individual devices and their particular applications are the main focus. Procedures have grown more complicated and interrelated to enable seamless interaction between devices, people, processes, and data. Data processing, analytics, decision-making, and automation are examples of advanced processes crucial for streamlining operations, improving consumer experiences, and developing new business models. Early linked devices could only generate a certain amount, kind, and complexity of data. The emphasis was gathering and storing information for particular applications or monitoring needs. From a variety of sources, such as sensors, devices, and user interactions, the IoE ecosystem generates enormous amounts of structured and unstructured data. Advanced data processing and analytics technologies are necessary to convert this data into usable information because it is essential for driving insights, decision-making, and automation. Wired connections or short-range wireless technologies like Bluetooth are initially the only connectivity choices for linked devices. These communication techniques limited the reach and expandability of IoT applications. Wi-Fi, cellular networks, LPWAN, and 5G are just a few of the communication technologies that are now widely available, which have increased the ecosystem's potential and reach. These technologies enable seamless data transfer and sharing by establishing faster, more dependable, and more effective connections between objects, people, and processes.

8.2 Introduction to Blockchain

Blockchain technology is a distributed database that enables users to store, trade, and use data traceably. It was revealed as the foundation for Bitcoin [1] in 2008 and was viewed as electronic money not controlled by centralized entities such as banks. Before Bitcoin, many decentralized protocols were established, such as BitTorrent [2], allowing users to share files without a centralized server. However, none of these attempts successfully resolved the issue of decentralized consensus. Bitcoin marked a significant advancement in decentralized technology, allowing users to make payments without the help of centralized banks and financial organizations [3]. As more people utilized Bitcoin, demand for blockchain technology grew.

How data is stored and delivered has changed due to this powerful technology. It is nearly impossible to hack or alter the data contained in a blockchain because each block's cryptographic hash creates a tamper-proof data chain [4]. Any changes to the data would need all network participants' consent to maintain the data's accuracy and validity. A cryptographic hash links each block in a blockchain to the one before it, and a blockchain is made up of many such blocks. A new block is added to the chain, and its contents are confirmed using a consensus procedure before it is included. This process ensures that all network users have agreed upon its contents.

The most common consensus mechanisms are proof of work and proof of stake [5]. A block becomes an immutable component of the blockchain when it is introduced to the chain and cannot be altered or withdrawn. Blockchain technology is the best option for situations when preserving data integrity which is essential because of the security and immutability it offers.

Thus, blockchain can be defined as a database type, more precisely, a distributed ledger database. Nonetheless, it is distinct from conventional databases in several essential aspects. Firstly, unlike traditional databases, which are frequently centralized and run by a single authority, blockchain databases are distributed throughout a network of nodes and are decentralized [6, 7]. This shows that the data is accessible to everyone online and that no single institution controls the blockchain database. Secondly, blockchain databases contain data in blocks connected in a tamper-proof data chain as opposed to traditional databases, which store data in tables. Thirdly, blockchain databases make it challenging to alter or manipulate the data by using cryptographic techniques to construct a tamper-proof data chain. This is not the case in traditional databases where different security controls, such as defining an access control list and using an intrusion prevention system, are required to protect the data. Lastly, blockchain databases are considered to be suitable and traceable, which means that any user can observe and verify them in real time, which is not the case in traditional databases [8].

8.3 Types of Blockchain

Before using blockchain technology, it is essential to understand the different types of blockchains. This will allow the selection of the best suitable blockchain type that will fulfill the requirements of the use case. Each type of blockchain has unique features, benefits, drawbacks, and limitations. By understanding these differences, we can choose the blockchain that best suits the application's needs. The blockchain types include public, private, consortium, and hybrid blockchains, as shown in Fig. 8.1.

Public blockchains are completely decentralized as they are open-source and allow anyone to manage the network making it completely permissionless [9]. People or organizations are encouraged to join the network by offering incentives like cryptocurrencies. Public blockchains are the crucial component behind the cryptocurrency ecosystem as they allow digital assets creation and transfer securely without any intermediaries. Every transaction is recorded in a public ledger available to anybody with an Internet connection, making them transparent and accountable. A robust developer community constantly working to improve the technology helps build user trust and confidence in public blockchains.

Due to the decentralized nature of public blockchains, they have several advantages but face challenges that may limit their use. The public blockchain can be expensive as the users must pay the transaction fee to execute their transactions on the network. This fee incentivizes the nodes that execute the transaction over the

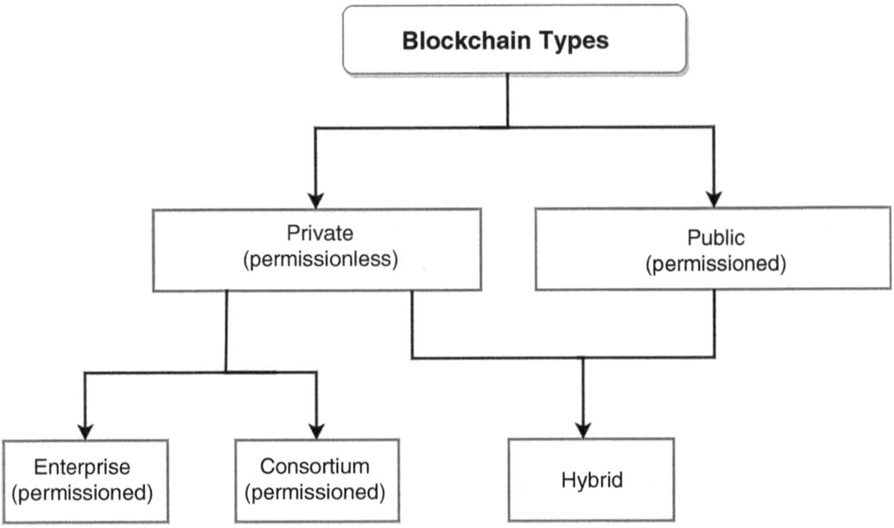

Fig. 8.1 Types of blockchain

network. The transaction cost is a critical obstacle in the selection of public blockchains. Another issue is scalability which is the network capacity to manage many transactions. Public blockchains can face slow transaction times and higher costs as the network gets congested. To make public blockchains more dependable, the cost and delay in executing a transaction must be reduced. Other than these two issues, the research community is aiming to address scalability issues. The two most well-known public blockchains are Ethereum and Bitcoin.

In contrast to Ethereum [10], which was created as a platform for creating decentralized apps utilizing smart contracts, Bitcoin is based on a proof-of-work consensus method. Programmers may create decentralized apps on top of the blockchain using Ethereum's smart contract capability; however, owing to its more complex consensus mechanism and use of gas prices to prioritize transactions, the two blockchains' transaction processing speeds differ. Compared to Bitcoin, Ethereum can process more transactions per second.

Private blockchains are controlled by a single or a limited number of nodes [11]. This gives private blockchains improved privacy and security and faster transaction speeds than public blockchains. Private blockchains only allow authorized persons to access the network, making them permissioned blockchains. Private blockchains provide the network owner control over who is allowed to participate and how the network's rules are implemented. Additionally, private blockchains give greater security and anonymity than public blockchains. Transactions may be completed faster since the network is smaller and there are fewer users than on a public blockchain. Faster transactions are critical for financial, healthcare, and supply chains. Private blockchains also have several disadvantages when compared to public blockchains. Private blockchains are only maintained by a single node or group of

Table 8.1 Comparison between private and public blockchains

Type of blockchain	Private	Public
Access	Permissionless	Permissioned
Advantages	Performance	Traceability
	Privacy	Transparency
	Faster transactions	Accountability
	Access control	Open source
Disadvantages	Semi-decentralized	Network delay
		Scalability
	Less transparent	Privacy
	Lack interoperability	Transaction fee
Use cases	Cryptocurrencies	Supply chain
Examples	Bitcoin	Hyperledger fabric
	Ethereum	R3 Corda

selected nodes, so they cannot be considered decentralized. The users must trust the nodes to provide secure transactions. This makes the blockchain less reliable and transparent. As private blockchains are permissioned, they also lack interoperability, making them difficult to connect to different systems. Therefore, businesses should consider whether a private blockchain is the best solution for their needs and whether a public blockchain would be a better match. R3 Corda [12] and Hyperledger Fabric [13] are famous examples of private or consortium blockchains. R3 Corda is a permissioned blockchain network designed mainly for financial institutions, providing smart contract functionality, privacy, and scalability. Hyperledger Fabric is an open-source blockchain framework hosted by the Linux Foundation, providing a modular and flexible platform to support applications in various use cases. Corda and Fabric have different architectures, with Fabric using a modular design and Corda using a distinctive architecture built on individual nodes that speak directly to one another. Table 8.1 shows the comparison of private and public blockchain.

Table 8.1 compares private and public blockchains to determine the appropriate blockchain technology for various applications. A third type of blockchain, hybrid blockchains, can be permissioned or permissionless. Hybrid blockchains allow businesses to control particular blockchain functionalities while retaining a public blockchain's security and openness. They are also more scalable than public blockchains, as the private part of the blockchain only needs a small amount of processing power to run. Ripple [14] is an example of a hybrid blockchain that combines elements of both public and private blockchains to allow speedier and more affordable cross-border payments. The Ripple network acts as a blockchain that is partially accessible to the public, allowing anybody to read transaction history and providing financial institutions with private channels to conduct transactions in confidence and secrecy. It can execute cross-border transactions in seconds instead of traditional systems and leverages a network of trustworthy validators to verify transactions. Ripple's hybrid blockchain allows organizations to benefit from a public blockchain's security and transparency while preserving control over sensitive data through private channels.

8.4 Understanding How Blockchain Works

Developing applications on blockchain platforms necessitates a detailed grasp of how transactions are processed. Understanding how blockchain works allows solution architects, engineers, and developers to construct safe, decentralized, and open solutions that use blockchain's unique properties, furthermore understanding blockchain facilities in selecting the ideal platform, ensuring security, and designing decentralized applications.

Blockchain transactions are exchanges of digital assets between two or more network participants using the sender, recipient, and value transfer type details. A block is a group of transactions recorded and verified as a single unit forming a blockchain structure. Each block contains a fixed amount of transactions and is broadcasted to all nodes in the network to validate its contents. It also has a unique digital signature generated using the set of transactions in the block plus the previous block's hash, allowing it to refer to the block before creating a secure chain of blocks. The genesis block is the first block in a blockchain, which defines the network's initial state and serves as the foundation for all succeeding blocks. It is typically generated by the blockchain's developer and is essential to the operation of the blockchain. Figure 8.2 illustrates transactions and blocks and genesis blocks and how they are connected to form a blockchain.

A node is a hardware or computer linked to the blockchain network that keeps a copy of the blockchain database. They are essential for the network's integrity, decentralization, and consensus rules, as they validate transactions, keep an accurate copy of the ledger, and enable data transfer. Consensus is a method blockchain networks use to ensure that all nodes have a current and correct copy of the database and that new transactions and blocks are added securely and transparently. To keep the blockchain network consistent, all participating nodes must agree on the same version of the blockchain. This is accomplished through the use of consensus-building approaches such as proof of stake and proof of work. Upon the creation of a block, all nodes in the network independently verify its information and agree to

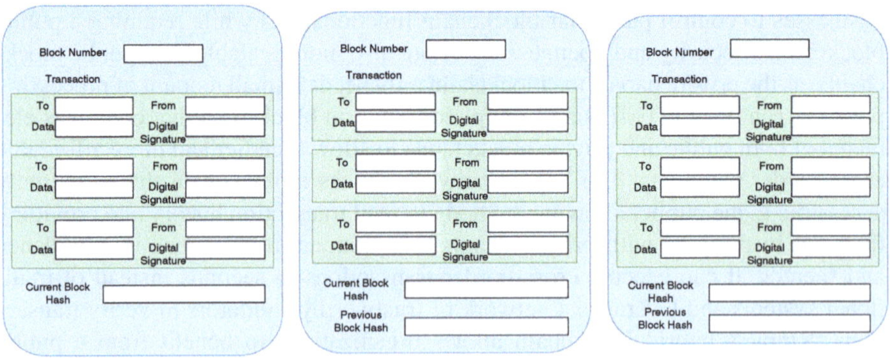

Fig. 8.2 Transaction, block, and chain of blocks

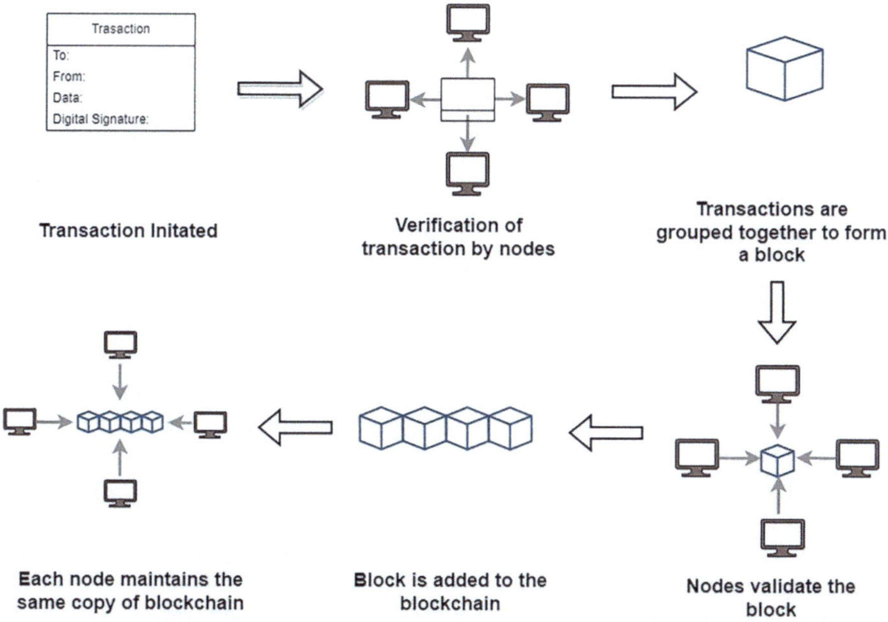

Fig. 8.3 Transaction initiation to completion in blockchain

its authenticity via consensus. When a public key is hashed, it produces an address, a unique identifier used to sign transactions and establish ownership of digital assets. Transactions may be received at public addresses as well.

Therefore, the complete process of the blockchain, starting from the transaction initiation to confirmation, is presented in Fig. 8.3 and further explained below:

- *Initiation of a transaction*: A user creates a digital signature with their private key to initiate a transaction. The transaction typically consists of the sender address, recipient address, and relevant data. As a digital signature protects the transaction, no one can pretend to be the user or alter the transaction data. The sender's public key confirms the digital signature's authenticity.
- *Verification by nodes*: After the transaction is initiated, it is broadcasted to the networks so that the nodes running the identical blockchain can validate and verify it. The nodes comply with the consensus mechanism to ensure the transaction's authenticity. Some blockchain protocols, such as proof of work, encourage nodes to compete for transaction validation by solving complex mathematical puzzles. Other systems, like proof of stake, select nodes based on how many tokens they presently hold.
- *Selection of transactions*: When the transaction has been validated, miners choose a group of transactions from the pool of unconfirmed transactions and add them to a block. By earning a transaction fee for each transaction included, miners are incentivized to include as many transactions as the block size allows.

Several factors, including transaction fees and size, may influence the selection process.
- *Addition to the blockchain*: When a block is approved, it is put to the blockchain and given a special code or hash that links it to the block before it. Each block's hash code keeps a tamper-proof, time-stamped record of all transactions. The blockchain's security and immutability are ensured by the hash function's unidirectional nature, which makes it simple to produce a hash code from data but challenging to generate data from a hash code.
- *Distribution and replication*: In the blockchain, every node must have the same copy of the database; therefore, it is distributed across the system. It is nearly impossible to hack or modify the blockchain since doing so would require the majority of nodes in the network to act maliciously. The distribution and replication operations of the blockchain ensure its decentralization and transparency.
- *Confirmation of transactions*: A transaction is confirmed when added to the block and distributed across the network, so each node has an identical ledger copy. Depending on the blockchain technology, confirmation time might range from a few minutes to many hours. Furthermore, the confirmation time can be affected by transaction cost, the number of nodes verifying the transaction, and the degree of network congestion. After a transaction is validated, it is permanently uploaded to the blockchain and cannot be modified, reversed, or withdrawn, making it safe and tamper-proof.

8.5 Role of Blockchain in the Internet of Everything

The IoE enables real-time communication and information sharing by connecting many devices, sensors, and systems to the Internet [15]. Yet, because of the enormous amount of data collected, it is crucial to ensure data security and privacy. In particular, device and network security problems might endanger important data and user information. IoE devices can gather private information about a person's whereabouts, health, and behavior. As a result, protecting data privacy and abiding by data protection rules are essential for upholding user confidence and avoiding legal ramifications. Strong security mechanisms must be implemented to safeguard data from unwanted access, interception, and alteration to address these problems. Data encryption and authentication methods can also provide another level of security to guarantee data privacy. Several steps may be taken to protect user and data security as well as to reduce the hazards connected to the Internet of Everything.

On the other side, blockchain technology offers a decentralized, unchangeable, and transparent way to store and distribute data while guaranteeing its security and integrity. One of the primary advantages of using blockchain technology for data storage in IoE is the absence of a central authority or intermediary. Without the control of central authority, the likelihood of data loss or corruption is reduced. Also, the implementation of blockchain technology ensures the legitimacy and validity of data storage by offering an unchangeable safe mechanism, which is essential in the IoE environment where information from multiple sources must be trusted. Blockchain technology may be used by businesses and governments to confirm the reliability, accuracy, and integrity of their data. In addition, blockchain technology may enhance privacy and anonymity, another benefit of using it for IoE data storage. Blockchain technology allows users to control their data and choose who can access it, ensuring their information is not used against their will. It may also be utilized for safe data exchange in the IoE in addition to data storage. Data may be purchased and exchanged safely and transparently between parties thanks to decentralized data marketplaces made possible by blockchain technology. Smart contracts may also be used to enforce access restrictions and data usage guidelines, guaranteeing that data is only shared with authorized parties and only for designated reasons.

Another major challenge of the IoE is interoperability. The devices and systems commonly use many data formats and protocols, and integrating and evaluating data from multiple sources may become problematic. Blockchain technology can help address the IoE interoperability challenge by providing a decentralized and secure channel for data transmission and integration across multiple devices and systems. Blockchain-based solutions can provide an industry-standard mechanism for IoE devices to share data, improving interoperability and cutting integration costs. Blockchain-based solutions have the potential to provide an industry-standard way for IoE devices to communicate data, enhancing interoperability and lowering integration costs. By simplifying data integration and sharing among devices and apps, the IoE ecosystem may operate more effectively and efficiently. Smart contracts may be used to impose data format standards, guaranteeing that data is organized consistently across all hardware and software. Smart contracts can automate the execution of corporate rules and procedures, while a distributed ledger can store and communicate data among many devices and systems. This might make it possible for gadgets and systems to work together without human involvement.

Let's use the healthcare industry to demonstrate how blockchain technology may be used for secure data exchange and storage in the IoE. Sensitive patient data is generated by many different technologies and systems in a healthcare system, including medical devices and health monitoring systems. This information should only be accessible to authorized individuals, including academics, insurance providers, and healthcare practitioners, and it must be handled securely. In this circumstance, blockchain technology might be employed in healthcare to provide safe data transfer and preservation. Furthermore, smart contracts might be utilized in the healthcare industry to enable secure data transmission between authorized parties. In addition, these agreements might be used to enforce access restrictions and data usage regulations, ensuring that data is only shared with authorized parties and for particular reasons. A healthcare provider, for example, may be granted access to a patient's electronic health record only if the patient has given permission.

By utilizing pre-defined rules and circumstances, blockchain technology can automate transactions between connected devices in the IoE, ultimately reducing the need for manual intervention and streamlining processes. Smart contracts can be used in the IoE to automate the execution of business rules and procedures between devices. Smart contracts, for example, can be used to automate the transfer of goods ownership between two devices when the conditions of the agreement are met, such as when payment is received and ownership of the product is transferred to the buyer instantly. Complex business processes in the IoE may be done automatically by utilizing smart contracts. For example, a smart contract might be created to automate the supply chain management process. When the contract requirements are met, the contract may be built to make payments and other activities automatically and track the flow of products from the manufacturer to the distributor to the retailer. The ability of smart contracts to make transactions automatically when specific criteria are met enables standardized and secure device communication. Because connected devices automatically conduct transactions and business processes, adopting smart contracts in the IoE may increase efficiency, reduce costs, and improve security.

Finally, blockchain technology can also be used to authenticate and identify IoE devices. Digital certificates, public and private key pairs, and decentralized identity systems may all be used with blockchain technology to verify and identify IoE devices. These techniques offer a safe and impenetrable means of confirming the identification of IoE devices, preventing illegal access and guaranteeing the network's integrity. Digital certificates can be kept on a blockchain and used as a tamper-proof, decentralized authentication technique in the IoE. A distinct digital certificate kept on the blockchain and given to each device is possible. Two devices can exchange digital certificates for creating a secure connection when they need to speak with one another. Another way blockchain technology may be used to identify and authenticate IoE devices is to employ public and private keys. Each device may be assigned a unique public and private key pair that is stored on the blockchain. Although the private key is used to sign transactions and establish secure connections, the public key may be used to validate the identity of a device. Finally, blockchain technology may be used to construct decentralized IoE device IDs. Each gadget may be assigned a unique identifier that is stored on the blockchain. Identification might include the details of the device such as the manufacturer or the machine identification number. This information can be used to authenticate the device's identity and ensure it has the authorization to access the network.

8.6 A Framework for Blockchain in IoE

A framework is required for integrating blockchain technology into the IoE since it provides a systematic technique for building and deploying a blockchain-based system. The IoE generates massive amounts of data that traditional data management systems are not designed to handle. Blockchain technology provides a decentralized, secure, and transparent platform for storing, sharing, and retrieving data; nevertheless, a framework is required to ensure that a blockchain-based system for the IoE is conceived, developed, and implemented in an organized and logical manner. The framework assists in establishing the IoE's data requirements, the appropriate blockchain platform to utilize, the consensus technique, and the access limitations and regulations to implement. Furthermore, the framework ensures that the blockchain-based system is integrated with the business's existing systems and procedures, ensuring the technology provides actual value to the organization. Some critical factors that might serve as the foundation of a framework for blockchain in IoE applications include the following:

1. *Define the use case*: Finding the exact use case of the IoE system is the first step in determining how blockchain technology can be applied in the IoE environment. It is necessary to understand processes, tools, and hardware used to generate data and the essential data properties that must be gathered, stored, and transferred while considering the constraints of the IoE environment. Defining the use case helps to identify the most suitable solution for a problem, determine the scope and boundaries of the solution, identify potential challenges and limitations, and set realistic expectations for the solution.
2. *Define the system requirements*: Establishing requirements is essential in creating a framework for integrating blockchain technology in IoE applications. This stage entails identifying the major stakeholders in the IoE ecosystem, such as device makers, network providers, end users, and regulatory authorities and stakeholders. Other variables must also be considered when determining the requirements, including scalability, performance, and cost-effectiveness. The development team can concentrate on adding the features and functionality the solution needs by outlining the requirements in advance. This helps to guarantee that the solution is fit for purpose and satisfies the stakeholders' expectations.
3. *Selecting the type of blockchain*: Prior to developing the system architecture, it is essential to select the type of blockchain. The choice of blockchain depends on a number of factors, including the level of decentralization required, the need for privacy and security, and the scalability and speed requirements of the application. Each type of blockchain, including public, private, permissioned, and hybrid, has strengths and weaknesses. By selecting the kind of blockchain early in the development phase, the solution may be tailored to meet the particular expectations of the stakeholders and designed to work within the restrictions of the chosen blockchain.
4. *Selection of consensus mechanism*: It is crucial to select the most appropriate consensus methods suitable for the particular use case. Security, scalability, performance, and energy efficiency trade-offs for various consensus algorithms vary. Proof of stake (PoS) is a more recent consensus technique that depends on validators owning a stake in the blockchain to confirm transactions and add blocks to the blockchain. Proof of work (PoW) is a frequently utilized consensus mechanism. A blockchain platform's consensus method is vital, and various blockchain platforms provide different consensus algorithms. While choosing a consensus mechanism, examining the use case's unique needs, such as transaction throughput, security, and energy efficiency, is critical. The correct consensus method and blockchain platform may be chosen based on these needs.

Blockchain-based IoE systems may be created to be safe, scalable, and effective by choosing the most suitable blockchain platform and consensus mechanism.

5. *Implement access controls and data usage policies*: In blockchain-based IoE systems, smart contracts may be utilized to define access restrictions and data usage regulations. Various blockchain platforms offer different smart contract languages. The complexity of the needed smart contracts and the programming language competence of the development team should be considered while selecting the smart contract language. Smart contracts can provide access restrictions to set unique roles or permissions for distinct users or devices. Data usage policies may be implemented by outlining how various parties can access and utilize data. Smart contracts may also be used to govern the use of personal data gathered by IoE devices, such as restricting data usage to certain objectives or obtaining authorization from the data subject before sharing data with other parties. Addressing data privacy and security concerns in the IoE environment while complying with legal and regulatory standards can aid in protecting sensitive information.
6. *Selecting data storage*: For IoE systems, choosing the best data storage solution is crucial. Data can be stored on-chain or off-chain. On-chain storage is secure and imputable, but it can experience issues with storage capacity, privacy, and transaction costs. On the other hand, off-chain storage is more adaptable and scalable and provides quicker transaction rates, albeit it may be less secure owing to the weaknesses and vulnerability of the underlying storage systems to hacking or data loss. Off-chain storage is frequently chosen over on-chain storage when privacy is an issue since it offers better data management and scalability. The suitable storage choice should be chosen based on the unique needs of the IoE application. Both storage methods can be used complementary to maximize security, speed, flexibility, and privacy, depending on the nature of the application and data.
7. *Governance and regulation*: Governance and legal considerations must be considered while developing an IoE application employing blockchain technology. Building governance structures, developing legal and regulatory frameworks, and carrying out network development plans are critical to ensure compliance with applicable rules and regulations. In developing an efficient governance structure, it is vital to consider the potential data protection, cybersecurity, and industry-specific regulation components of legal frameworks. Such a system should incorporate network management processes, legislation, and regulations. The network's borders must also be defined, congestion management procedures developed, and rules put in place to deal with increasing traffic. These aspects should be considered while developing a blockchain-based system to ensure conformity to legal standards and effective network administration.

8.7 Conclusion

To attain IoE's full potential, data security, privacy, interoperability, and device authentication must be addressed. Blockchain technology may be used to solve these problems. For instance, blockchain technology may make automated processes, secure and decentralized data flow and storage, system and device interoperability, and device authentication possible. As a result, businesses, governments, and people involved in the IoE ecosystem may improve their security and privacy safeguards while attaining better levels of interoperability by using blockchain. Therefore, it is crucial to keep investigating the capabilities of blockchain to realize the IoE's full potential.

References

1. Bitcoin Whitepaper (2021). Interventions. https://doi.org/10.21428/9610ddb2.a6a2490c
2. Legout A, Urvoy-Keller G, Michiardi P (2005) Understanding BitTorrent: An experimental perspective.
3. Gregoriou GN, Nian LP (2015) Handbook of Digital Currency: Bitcoin, Innovation, Financial Instruments, and Big Data. The Journal of Wealth Management 18 (2):96–97. https://doi.org/10.3905/jwm.2015.18.2.096
4. Singhal B, Dhameja G, Panda PS (2018) How Blockchain Works. Beginning Blockchain. Apress. https://doi.org/10.1007/978-1-4842-3444-0_2
5. Lashkari B, Musilek P (2021) A Comprehensive Review of Blockchain Consensus Mechanisms. IEEE Access 9:43620–43652. https://doi.org/10.1109/access.2021.3065880
6. Javed IT, Qureshi KN, Alharbi F, Jeon G (2022) Terahertz fading model for wireless nanosensor networks in advanced medical manufacturing technologies. The International Journal of Advanced Manufacturing Technology. https://doi.org/10.1007/s00170-022-09660-9
7. Qureshi KN, Shahzad L, Abdelmaboud A, Elfadil Eisa TA, Alamri B, Javed IT, Al-Dhaqm A, Crespi N (2022) A Blockchain-Based Efficient, Secure and Anonymous Conditional Privacy-Preserving and Authentication Scheme for the Internet of Vehicles. Applied Sciences 12 (1):476
8. Qureshi KN, Jeon G, Piccialli F (2020) Anomaly detection and trust authority in artificial intelligence and cloud computing. Computer Networks:107647
9. Andreev RA, Andreeva PA, Krotov LN, Krotova EL (2018) Review of Blockchain Technology: Types of Blockchain and Their Application. Intellekt Sist Proizv 16(1):11. https://doi.org/10.22213/2410-9304-2018-1-11-14
10. Buterin V (2014) A next-generation smart contract and decentralized application platform. white paper 3(37):2–1
11. Pahlajani S, Kshirsagar A, Pachghare V (2019) Survey on Private Blockchain Consensus Algorithms. Paper presented at the 2019 1st International Conference on Innovations in Information and Communication Technology (ICIICT), 2019/04
12. Brown RG (2018) The corda platform: An introduction. Retrieved 27:2018
13. Androulaki E, Barger A, Bortnikov V, Cachin C, Christidis K, De Caro A, Enyeart D, Ferris C, Laventman G, Manevich Y, Muralidharan S, Murthy C, Nguyen B, Sethi M, Singh G, Smith K, Sorniotti A, Stathakopoulou C, Vukolić M, Cocco SW, Yellick J (2018) Hyperledger Fabric. Paper presented at the Proceedings of the Thirteenth EuroSys Conference, 2018/04/23
14. Todd P (2015) Ripple protocol consensus algorithm review. May 11th
15. DeNardis L (2020) The Internet in Everything. Yale University Press. https://doi.org/10.12987/yale/9780300233070.001.0001

Chapter 9
Cybersecurity as a Service

John Morris, Stefan Tatschner, Michael P. Heinl, Patrizia Heinl, Thomas Newe, and Sven Plaga

9.1 Introduction

Cybersecurity as a service (CSaaS), also sometimes referred to as Security as a Service (SECaaS) [1], is the outsourcing of key IT security functions to an external specialist company or third party. The concept of CSaaS ultimately began back in 1987 with the availability of the first antivirus product called VirusScan from McAfee [2] where computer users paid to be protected from malware attacks. Roll on 30 years and as the malware has become more abundant and complex, the need for more protective services has increased in tandem. The initial uptake on this new breed of cybersecurity services with names like vulnerability assessment and chief

John Morris and Stefan Tatschner contributed equally to this work.

J. Morris
Department of Electronic and Computer Engineering, University of Limerick (UL), Limerick, Ireland
e-mail: john.morris@ul.ie

S. Tatschner (✉)
Department of Electronic and Computer Engineering, University of Limerick (UL), Limerick, Ireland

Department Product Protection and Industrial Security, Fraunhofer AISEC, Garching bei München, Germany

Confirm, the SFI Centre for Smart Manufacturing, Limerick, Ireland
e-mail: stefan.tatschner@aisec.fraunhofer.de

M. P. Heinl
Department Product Protection and Industrial Security, Fraunhofer AISEC, Garching bei München, Germany
e-mail: michael.heinl@aisec.fraunhofer.de

K. Naseer Qureshi et al. (eds.), *Cybersecurity Vigilance and Security Engineering of Internet of Everything*, Internet of Things,
https://doi.org/10.1007/978-3-031-45162-1_9

information security officer (CISO) is a service has been passive. One cause for this slow engagement is that many chief executive officers (CEOs) believed investment in such services is an unnecessary expense. On the technical side, some IT directors feel that their positions within the company structure are endangered and they are confident that they can do it better themselves, anyway, particularly in the case where the outsourcing of key organizational security functions to outside contractors is concerned.

The recent increases in cyberattacks of high-profile companies around the world [3] and better cybersecurity education have altered this mindset in a positive way. Additionally, it has been proven that most organizations are still reactive when it comes to cybersecurity. They still believe that a malware attack will not happen to them: so why pay for cybersecurity? It is deemed too high a price for embracing the concept of precaution. However, when such deniers are stroke by a sudden malware attack, suffering untold data losses or paying ransoms to the cybercrime-as-a-service industry, these entities suffer greatly for their negligence. That is, if they are still even in business after the attack, currently over half of all small businesses close within 6 months of a malware attack [4].

What is for certain though is that the volume of malware attacks is set to increase and become more sophisticated, particularly with the advent of malware enhanced by artificial intelligence (AI) like DeepLocker [5], and few companies will have the expertise and resources to deal with this evolving cyber-problem. Another point of note is that the malware attack surface is no longer confined to large networks of connected computers and servers, poorly written web interfaces, and e-mail phishing attacks. The newer malware is targeting the entire Internet of Everything (IoE) landscape. From mobile phones to smart wearables and resource-constrained Internet of Things (IoT) devices to cloud-based platforms. With such a large IT ecosystem to protect, it has become increasingly expensive for companies to train their IT staff to protect this attack surface or hire dedicated IT security staff. This is compounded by the fact that there is currently a worldwide shortage of IT security staff with current estimates at 3.4 million vacant positions [6].

CSaaS appears to be a step in the right direction to handling this growing threat landscape and allows companies to pick the IT security functions that they most need help with at a more affordable monthly rate. Simultaneously, not least due to

P. Heinl
Technische Hochschule Ingolstadt, Ingolstadt, Germany
e-mail: patrizia.heinl@aisec.fraunhofer.de

T. Newe
Department of Electronic and Computer Engineering, University of Limerick (UL), Limerick, Ireland

Confirm, the SFI Centre for Smart Manufacturing, Limerick, Ireland
e-mail: thomas.newe@ul.ie

S. Plaga
Center for Intelligence and Security Studies (CISS), Neubiberg, Neubiberg, Germany
e-mail: sven.plaga@unibw.de

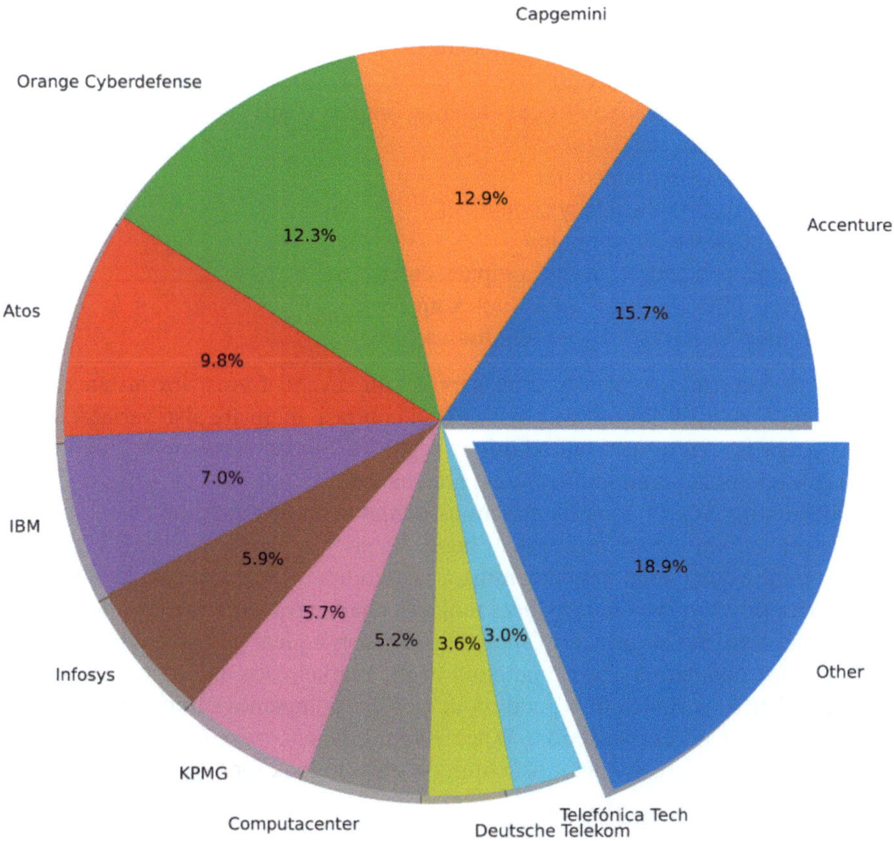

Fig. 9.1 Managed and professional security services market: revenue share of top participants, Europe, 2022, conducted by Frost and Sullivan [7]

the rising numbers of supply chain attacks, it is important that a provider is chosen who does not only offer an increase in security to its customers just from a technical viewpoint. To be able to protect sensitive customer data, a strong security ethos is also required on the provided services. Over the course of this chapter, a more in-depth review of the most common IT security functions being offered by CSaaS companies will be discussed. Also, a comparison of the main CSaaS companies will be conducted. Finally, a checklist will be created for companies looking to choose a CSaaS for themselves. Figure 9.1 shows the managed and professional security services market: revenue share of top participants, Europe, 2022, conducted by Frost and Sullivan.

The cybersecurity market has developed into one of the most profitable IT markets over the last decade [8]. Consequently, a lot of new IT companies specialized in cybersecurity were only founded in recent years or where existing IT companies launched dedicated cybersecurity divisions. According to the revenue study shown in Fig. 9.1, the top ten companies in the managed and professional security services market in Europe are the following:

- Accenture (https://www.accenture.com)
- Capgemini (https://www.capgemini.com)
- Orange (https://orange.com)
- Cyberdefense (https://www.cyberdefensecompany.com)
- Atos (https://atos.net)
- IBM (https://www.ibm.com)
- Infosys (https://www.infosys.com)
- KPMG (https://www.kpmg.us)
- Computacenter (https://www.computacenter.com)
- Deutsche Telekom (https://www.telekom.com)
- Telefonica Tech (https://www.telefonica.com)

CSaaS companies typically offer services in several forms, for instance, subscription or payment for utilized services. In contrast, there are also variants where basic usage is free to use, but additions (e.g., 24/7 customer support, higher rate limits, or additional premium features) are charged.

Outsourcing key IT security functions comes with benefits like cost cutting, a consistent and unified architecture, or better security expertise (by the CSaaS company). On the other hand, implementing CSaaS relies on sensible data being sent to the service provider which introduces multiple challenges requiring a well-designed architecture to avoid insecure applications. Consequently, companies offering CSaaS must maintain a good reputation in the marketplace and be trusted to stay relevant. The importance of a good reputation for companies offering CSaaS begs the question of decent selection. When looking to choose a CSaaS company to engage with, what are the ten most common traits to look for?

1. *How long is the entity in business?*
   The reputation is easier to spot when the entity is in business for a long time. In this case there may be online reviews, news articles, or similar material from third parties available.
2. *What companies is the entity working with already?*
   Collaborating with big players in the same area of work can be a hint for a good and trusted reputation, particularly, it these are long-term customers.
3. *What range of services does the entity offer?*
   Offering few services could be a hint for a highly specialized entity offering high-quality services. Are the specific services that are being sought being offered by the entity?
4. *What kind of service delivery model is employed?*
   On-premise, remote, or both? This trait is very specific to the relevant use case and the current security posture of the client company. On-premise means that dedicated resources and staff need to be provided, but a certain level of control is still ensured.
5. *Is it a fully managed service, or do internal IT resources have to be dedicated to delivering the services?*

This depends on whether the client company has internal staff with the requisite skills and time to manage the security requirements of the company. Fully managed is designed for client companies with little or no internal security personnel or systems.

6. *What type of pricing model is offered?*
   Fixed monthly, annually, per employee or device? This will depend on the type of security service being offered. Security training is typically charged by employee, whereas penetration testing and cyber-insurance can be charged monthly or annually.
7. *What is the skillset and qualifications of the staff?*
   Are the staff certified or doing public speeches at conferences in their area of work? Is their training relevant and kept up to date? Where are the gaps in the security staff skills that need to be filled by an external security company?
8. *Has the entity published any articles, or does the entity take part in any blogs or forums in the areas of cybersecurity?*
   This is a big indication of a security company that is highly skilled and extremely competent. It also means that they are keeping up to date with the latest security threats and trends.
9. *Does the entity provide a trial period or proof of concept?*
   This can be helpful in deciding if a particular security company or tools is compatible with the needs of a client company. A proof of concept can provide a try before you buy type scenario to help key decision-makers in the approval process.
10. *Is the entity certified?*
   Certifications help ensuring at least a minimum level of security. It also gives the client company a comfort in knowing that the entity has the requisite security qualifications to complete the security services being offered.

This book chapter contributes a list of ten most common traits to look for when choosing a CSaaS company. In addition to these traits, common CSaaS functions are researched and are related with high revenue companies. Furthermore, an overview over the current market share of professional CSaaS providers with a comparison about the offered services is given.

9.2 CSaaS Functions

The number of different cybersecurity services offered by these companies is substantial, especially when specialized use cases are included. However, the Cloud Security Alliance has published an overview [1] where a categorization of cybersecurity services was carried out. The provided categorization was enhanced by additional services based on our practical knowledge and logical reasoning. The identified key services are described in the following sections.

9.2.1 Security Personnel as a Service

CISO as a service or virtual CISO is the outsourcing of the chief information security officer role within an organization. This resource can work onsite within a particular organization or work remotely, reporting directly to the C-level group which is key for decision-making. They can work independently or as the head of a security team and work for a fixed contract period or month-to-month. Their duties include the following:

- Full review of an organization's security position.
- Recommend best practice hardware, software, and security changes. This can also include purchases.
- Interview, vet, and hire new security staff.
- Train internal security team.
- Generate penetration testing report.
- File NIST 800 security reports where required.

This role is more suited to mid- to large-sized companies where the budget for a permanent CISO role is currently not available or as a try before you buy type scenario. A main constraint of this approach is the often steep learning curve for the contractor in terms of corporate knowledge, cultural norms, and company politics. However, this last point can also be an advantage as the contract CISO is not affected by internal conflicts or job security.

Additional security roles that can be outsourced include a data protection officer, compliance and risk officer, forensic analyst, security trainer, penetration tester, and security helpdesk personnel.

9.2.2 Cyber-awareness Training

Cyber-awareness or malware threat detection training involves the systematic education of company employees in how to correctly identify malware threats, since 95% [9] of current company malware breaches are caused by human error. The format of the training is usually a step-by-step guide containing videos and a series of items to identify afterward, to reinforce the training. The training usually finishes with a quiz of all the topics discussed in the session with a completion certificate produced for a passing grade. The most popular cyber-awareness training programmers concentrate on e-mail phishing and social engineering attacks, in other words, training employees to think before clicking on that web link and entering their login credentials into a fake website like in Fig. 9.2. The training normally lasts around 30–40 min with some like the Kevin Mitnick-inspired KnowBe4 e-mail phishing offering lasting 50 min. The cyber-awareness training is then reinforced further with weekly mock phishing attacks being sent out to all employees. Training should be retaken by employees at least once a year to keep abreast of new types of malware

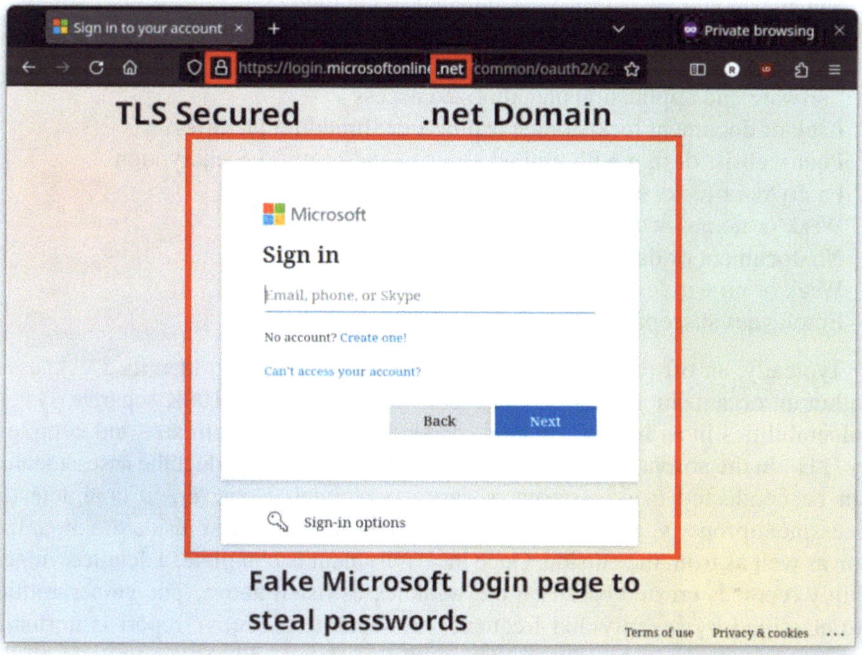

Fig. 9.2 Screenshot of a phishing website for a Microsoft login [10]

attacks. The training is offered as a managed service that typically reports to the Human Resources (HR) department rather than IT. The main types of cyber-awareness training sessions include the following:

- Phishing, smishing, and vishing attacks
- Remote work training
- General Data Protection Regulation (GDPR) training
- Foreign travel dos and don'ts
- Intellectual or physical company property training

The cyber-awareness training can also be customized with corporate branding and content to make it more realistic to the employees (e.g., actual company e-mails) and assist in the process of turning them into human firewalls.

9.2.3 *Vulnerability Assessment*

A vulnerability assessment involves the systematic identification, measurement, and categorization of weaknesses within an organization's systems. These weaknesses can take the following forms:

- Unpatched and unencrypted servers and/or computers
- Poorly setup firewall with open rules and port access
- Remote access vulnerabilities
- Software and application unauthorized access
- Lack of document lock storage cabinets or shredding facilities
- Poor website design with limited security and/or no TLS encryption
- Faulty door locks or doors left open
- Weak or no password policies
- No document or data audit process
- Weak or no wireless access point security
- Employees susceptible to social engineering attacks

Typically, an off-the-shelf vulnerability scanner is used to identify weaknesses within an organization. Current scanners can identify over 100K separate system vulnerabilities in as little as an hour, depending on the system size and complexity [11] . In the absence of in-house security personnel to conduct the assessment, it can be conducted using external security personnel. However, to complete the assessment properly, all systems will need to be scanned from inside the organization as well as from the outside. Once the assessment is complete, a detailed vulnerability report is created based on the weaknesses listed above. The vulnerabilities are classified by severity and frequency. A separate executive report is normally produced for the key decision-makers with less detail and more emphasis on the risks and financial impact to the organization.

9.2.4 Periodic Penetration Testing

Periodic penetration test is an authorized simulated cyberattack on a computer system, performed on a regular basis to evaluate the security of the system. Its objective is to identify vulnerabilities that could otherwise be used by malicious actors to abuse the computer system. A penetration test needs to be performed by a technical domain expert who can use similar techniques as those used by attackers. Penetration testing is a demanding task, and the following challenges apply:

1. *Staying up to date:* With the current state of the art from a technical standpoint. The IT sector is developing at a very fast pace, and a penetration tester must be capable of all the current and relevant technologies when conducting an effective test.
2. *Scope:* Defining the scope of testing is a challenging task. On the one hand, a scope that is too narrow might not yield useful results. On the other hand, too broad a scope could be unfeasible from a management perspective.
3. *Realistic attack scenarios* are considerable for a penetration test, since a highly academic attack scenario could indeed yield results. However, these results are at risk of not being relevant for the desired use case of the product.

4. *Limited access:* The integration of cybersecurity in the development process (i.e., security by design) is desired, since technical design decisions often have an impact on the security of a system. However, penetration testing during development can be restricted, since parts of the system might not be implemented yet.
5. *Reproducing issues:* Reproducing findings needs the careful documentation of all involved working steps and parameters of the test environment. Monitoring every relevant parameter in a penetration test is a difficult task, since all included parameters might not be known by the penetration tester at the offset.
6. *Time constraints:* Penetration testing is a complex task including creative components where good findings do not strongly correlate to the amount of time being spent on a test. However, budgeting in the first place can limit the effectiveness of penetration testing, since it limits the creativity of the tester.
7. *Collaboration and integration* with the development team is required for the feedback loop to integrate any findings improving the actual product.
8. *Skills:* Finally, the skillset of the penetration tester must be accurate for the relevant architecture and used technology.

Security by design is becoming more and more important in the design process of software products. Companies are beginning to integrate Secure Software Engineering into the relevant value chains [12]. Periodic penetration testing is a good option for evaluating that the designed software architecture is secure and that included security measures serve their purpose. However, in order to be effective, it requires careful planning and implementation.

9.2.5 E-mail Security

E-mail security is a critical component of an organization's communication. Due to its legacy, e-mail suffers from many design issues related to security. For instance, the content of an e-mail is usually only secured from the e-mail client to the e-mail server rather than being end-to-end secure. E-mail was designed at a time when the Internet was mainly an academic tool and thus end-end-security was not relevant. However, the success of e-mail especially in a corporate context might be a result of this simplicity. There are several key technologies available which are implemented by default by the common big e-mail service providers. Since e-mail does not provide any of these technologies by default, they were added on top, for example, adding metadata via e-mail headers.

1. *Encryption:* A procedure of converting plain text into a so-called cipher text, which can only be decrypted with a specific key. Encryption implements the protective goal of confidentiality both at transit and at rest. Most commonly used state-of-the-art technologies are Secure/Multipurpose Internet Mail Extensions (SMIME) or Pretty Good Privacy (PGP).
2. *Digital signatures:* Digital signatures are used to verify the authenticity and integrity of messages by using special metadata which is attached to a message.

In other words, these signatures can be used to verify that the message has not been tampered with during transit and that it was sent by the claimed sender. Most commonly used state-of-the-art technologies are Secure/Multipurpose Internet Mail Extensions (SMIME) or Pretty Good Privacy (PGP).
3. *Spam filters:* Filters which use sophisticated techniques to block unwanted messages.
4. *Anti-malware solutions:* Use signature-based detection, heuristics, or machine learning to identify and block messages that contain malware, such as viruses, Trojans, or spyware.
5. *Sender Policy Framework (SPF):* A protocol that allows organizations to specify which mail servers are authorized to send e-mails on their behalf.
6. *DomainKeys Identified Mail (DKIM):* A protocol that allows organizations to digitally sign e-mail messages on the server side to verify the authenticity and integrity of the message.
7. *Domain-Based Message Authentication, Reporting and Conformance (DMARC):* A protocol that allows organizations to protect their domains from unauthorized use, such as phishing and e-mail spoofing. DMARC allows organizations to publish policies that specify how recipient mail servers should handle e-mails that fail SPF and DKIM authentication.
8. *Authenticated Received Chain (ARC):* A protocol that provides a chain of authentication results for an e-mail message, starting from the original sending mail server to the recipient's mail server.
9. *Transport Layer Security (TLS):* A protocol that is used to provide communications security over a computer network. Due to its current widespread use in instant messaging, file transfers, and web traffic, TLS has become a basic technology for secure Internet today. Figure 9.3 shows the added header fields and the structural changes of an e-mail when ARC, DMARC, DKIM, SMIME, and SPF are in place.

```
From: sender_email_address
To: recipient_email_address
Subject: email_subject
MIME-Version: 1.0
ARC-Seal: arc_seal_value
ARC-Message-Signature: arc_signature_value
DKIM-Signature: dkim_signature_value
DMARC-Record: dmarc_record_value
Received-SPF: pass (sender_ip_addr: domain_of_sender designated_server_ip_addr permitted)
Authentication-Results: domain_name;
    spf=pass smtp.mailfrom=sender_email_address;
    dkim=pass header.i=@domain_name;
Content-Type: application/pkcs7-mime; smime-type=enveloped-data; name=smime.p7m
Content-Disposition: attachment; filename=smime.p7m
Content-Transfer-Encoding: base64


base64_encoded_SMIME_message_body
```
Structure of an e-mail with ARC, DMARC, DKIM, SMIME, and SPF

Fig. 9.3 E-mail structure

Due to this added complexity on top of the basic e-mail design, running a secure e-mail service is relatively cumbersome, especially as a violated or missing protocol could impair successful delivery of e-mails. Consequently, there are several companies that are specialized in providing secure e-mail services. Well-known free e-mail providers utilizing most of the mentioned key technologies are Google with its GMail[1] service and Microsoft with Exchange[2].

9.2.6 Identity and Access Management

Identity and Access Management (IAM) is a basic requirement of every effective security program in order to protect data, applications, and other assets. To be able to technically enforce it, i.e., only authorize legitimate requests, users must be reliably authenticated. This is usually done leveraging digital identities, e.g., usernames, which are linked to a person's actual identity. Typical standards used in this context are OAuth [13], OpenID [14], and Security Assertion Markup Language (SAML) [15]. Establishing and managing these digital identities seem to be a straightforward task but can become very complex once the number of employees and other stakeholders of an organization increases.

Therefore, IAM providers do not only offer the corresponding technologies but also best practices in the form of pre-defined processes and concepts. Typical functionalities offered by IAM providers include but are not limited to the following:

- Initial registration of users
- Assignment of roles and privileges
- Creation, provision, and management of credentials
- Centralized management of identities, roles, and privileges
- Centralized authentication and authorization of users
- Provision of means for multi-factor authentication (MFA)
- Support of interfaces for single sign-on (SSO) services

Accounts with a very high level of privileges, e.g., administrators or super users, are a popular target of threat actors and prone to insider risk. They should therefore be additionally protected leveraging privileged access management (PAM).

9.2.7 Cyber Insurance

In the last few years, the frequency and impact of cyber-incidents against companies worldwide continued to increase steadily [16]. While some industry segments were hit less frequently than others [17], there is no guarantee for anyone to be spared to

[1] https://gmail.com

[2] https://outlook.live.com

move into the focus of threat actors. Hence, no matter how much money a firm spends on its security program or which technical prevention controls it implements, there is a residual risk of being hit by a cyberattack that might lead to reputational and/or financial loss for the victim.

The purpose of cyber insurance is to step in if an insured victim experiences such a reputational or financial loss arising out of a covered cyber-incident. Coverages that are generally offered by insurance companies include the following:

- First-party damages (i.e., losses directly occurred to the policyholder) covering own costs (e.g., business interruption costs, incident response and forensics expenses, the launch of public relation campaigns, installation of call centers to inform customers).
- Third-party liability (e.g., claims made against the policyholder by a third party) covering costs to indemnify the claimants for a loss and the expenses of defending lawsuits associated with it. In many cases, these losses arise from the failure of an organization to appropriately protect third parties' data from being breached or compromised through a cyber-incident.

Additionally, many insurance carriers offer further services to their customers such as establishing connections to forensic and incident response firms as well as consultancy services. This is beneficial for both, the insurance carriers and the insureds, as both are interested in quick recovery after an incident to reduce costs. While the process for a company getting cyber-insurance certainly can differ, there are some steps each carrier performs before offering a binding quote for cyber-coverage:

1. Assessment of cyber exposure based on industry, company size, and business model
2. Evaluation of security protection level by on-site visits, conversations, questionnaires, and/or cyber-risk scanning and analytics tools
3. Legal wording of cover elements and exclusions
4. Actuarial calculation of potential losses, maximum capacity, and corresponding premium

With the recent surge of cyber-incidents, insurance companies started to be more selective on offering cyber-insurance. Companies need to fulfill minimum security standards defined by each carrier. In addition to that, insurers need to protect themselves from large-scale events which can hit multiple clients at once, so-called accumulation risks. Scenarios which are under discussion and currently excluded by most carriers are cyber-incidents which arise out of any kind of cyber-war (whether declared or not) and the outage of external networks, such as the Internet or electricity supply.

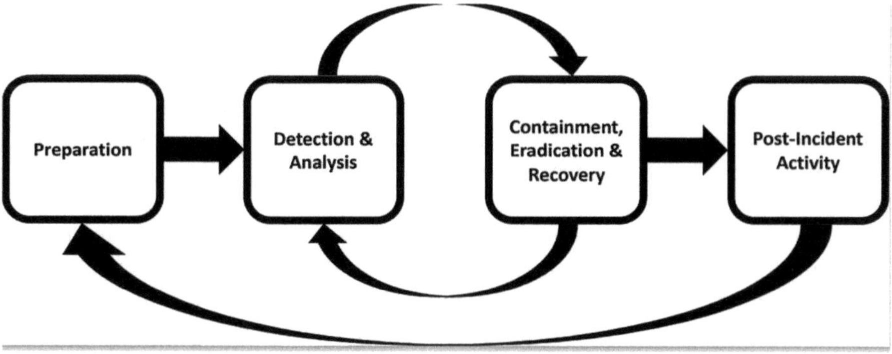

Fig. 9.4 Incident response life cycle according to NIST SP 800-61 Rev. 2 [18]

9.2.8 Incident Response

There is a saying that companies should not ask themselves if they are vulnerable to a security incident but only when and to which extent this incident may occur. Keeping that in mind, it is important to be prepared for the moment in which such an incident happens. Therefore, incident response (IR) services should not only provide support during an incident. According to the National Institute of Standards and Technology (NIST), the incident response life cycle encompasses a total of four phases as shown in Fig. 9.4:

1. Preparation
2. Detection and analysis
3. Containment, eradication, and recovery
4. Post-incident activity

Ideally, an IR service covers all of these phases. This makes rapid response much more likely, as information from all phases is directly available during the actual IR and does not have to be shared cross-organizationally among different service providers, which would cost valuable time.

Before the actual incident, incident response services encompass consultation on technology enabling the customer to detect and contain incidents, e.g., solutions for Security Information and Event Management (SIEM) and Endpoint Detection and Response (EDR). Furthermore, one of their technological focuses is on configuring the customers' infrastructure not only securely but in a way that retains and protects information which is valuable for incident handling and investigation, e.g., read-only backups and audit logs. Apart from these technological measures, IR also encompasses preparation on an organizational and human level, including the preparation of customized response plans and playbooks as well as regularly putting their content into practice through tabletop exercises. Ideally, these tabletop exercises are as inclusive as possible, involving not only representatives from IT (security) but also from operations, legal, human resources, public relations, etc.

For the case where a potential incident has been detected, IR services ideally offer an emergency hotline which can be consulted 24/7 in order to provide support during the process of triage and first response. Once it is confirmed that the initial alarm has not been a false positive, IR services begin with evidence collection and root cause analysis. In order to be prepared for potential court cases and to support law enforcement, it is paramount to document the analysis as thoroughly as possible and maintain the chain of custody during forensics.

When affected parts of systems and networks are identified, an appropriate containment strategy, such as powering them off or disconnecting them from other parts of the network, has to be chosen. The choice heavily depends on the pursued, sometimes conflicting objectives besides the actual containment, e.g., preserving evidence even in non-persistent memory or stopping a ransomware attack from continuing to encrypt data. Once the threat is contained, it has to be eradicated, e.g., by wiping malware, mitigating vulnerabilities, and disabling compromised accounts. After that, recovery can take place, e.g., by resetting passwords and restoring systems.

As indicated in Fig. 9.4, the described phases are not strictly linear but rather part of an iterative, recurring process. Depending on the organizational and technological environment of the individual incident, IR engagements can happen on premise, remotely, or in a mixed mode, depending on the phase

9.2.9 Business Continuity/Disaster Recovery Planning

The planning of IR and business continuity/disaster recovery (BCDR) is closely related. However, the scope of BCDR goes beyond potential business interruptions caused by security incidents and does primarily focus on the continuity and recovery of the core business, i.e., keeping critical processes running independently from the environment or restore them as quick as possible, respectively. Since these core processes change over time, BCDR also must dynamically adapt and is therefore not a task to do once but a continuous process which can be managed systematically according to ISO 22301. Just as IR, BCDR is a highly interdisciplinary process involving various stakeholder groups to discuss and define a desirable yet realistic recovery time objective (RTO), recovery point objective (RPO), as well as the corresponding measures. BCDR as a service can include the organizational part of moderation, consolidation, and documentation of these stakeholders' requirements in the form of a BCDR plan but also what is called recovery as a service, meaning backup and restore solutions hosted in the cloud.

9.2.10 Security Information and Event Management

As previously mentioned, SIEM can be very helpful when it comes to the detection and investigation of security incidents. Besides the pure aggregation of potentially security-related information, e.g., log files or real-time network data, from a variety of sources, it can also offer continuous monitoring and correlation to automatically (e.g., by anomaly detection) or semi-automatically (e.g., by pre-configured use cases) detect suspicious activities. Additional factors to be considered are intuitive user interfaces and flexible support of formats and protocols to include data from as many nodes as possible, as well as the scalability to be able to serve the dynamic landscape of a growing business. Apart from the option to deploy and use it on premise, it can also be deployed in the cloud and observed by well-trained analysts of the provider, ideally working in shifts to provide 24/7 coverage. This comes with the advantage that security alerts can be analyzed directly when they happen, i.e., without long delays after business hours or on weekends

9.2.11 System Patching and Updates

With the disclosure of software vulnerabilities, vendors are required to correct them as fast as possible, since they might be discovered and exploited by attackers to gain access to a computer system. Reacting as fast as possible to disclosed vulnerabilities is commonly called patching, since it is critical to pre-empt attackers. Good historical examples where software updates were mission critical are Heartbleed[3], Triple-Seven[4], Shellshock[5], and EternalBlue[6]. What these vulnerabilities have in common is a large and possibly fatal impact on the attacked IT infrastructure:

- They can be easily discovered by an attacker.
- They are easily exploitable (usually few lines of, e.g., Python code).
- They have a fatal impact, for instance, remote code execution (RCE) or sensitive information leaks.

Fortunately, software updates for such kinds of critical vulnerabilities usually are available very quickly. For instance, patches for the famous Heartbleed vulnerability were available even before it was privately disclosed to the development team. Seven days after the disclosure, an official release of the affected software was

[3] https://cve.mitre.org/cgi-bin/cvename.cgi?name=CVE-2014-0160

[4] https://cve.mitre.org/cgi-bin/cvename.cgi?name=CVE-2016-0777

[5] https://cve.mitre.org/cgi-bin/cvename.cgi?name=CVE-2014-6271

[6] https://cve.mitre.org/cgi-bin/cvename.cgi?name=CVE-2017-0143

available[7]. At the time of disclosure, there were around 300,000 vulnerable servers online. It is surprising that 6 years later there were still 200,000 vulnerable servers online[8].

These examples show the necessity of keeping up with evolving threats. Therefore, cybersecurity systems need to track the current state of the art of available countermeasures. For instance, software modules that process untrusted data are one of the most critical parts to protect, as they are directly accessible by attackers. Operating systems provide mechanisms offering basic protection which in general limit the attack surface. In order to benefit from such cautionary measures, regular security updates and reviews are desired.

Software updates in production are rolled out via well-established update mechanisms. In free and open-source software (FOSS) environments, packet management systems, such as apt, dnf, or pacman, are common. Usually, there are different update tracks including stable updates (i.e., stability and security updates) or bleeding edge (i.e., new features are deployed as fast as possible). In non-FOSS environments, there might be proprietary solutions with similar semantics. Careful reviews of the used software repositories are required when building products or infrastructures relying on these updates. CSaaS companies ensure that maintained components or services stay up to date and are not affected by known vulnerabilities.

9.2.12 Security Standards Compliance

With the rising number of networked devices and digitization of most parts of our lives in the context of the Internet of Everything, the number of security-related regulations and industry-specific standards which need to be considered continuously increases. Examples include the following:

- General Data Protection Regulation (GDPR)
- ISO/IEC 27001 Information Security
- ISA/IEC 62443 Cybersecurity for Operational Technology
- ISO/SAE 21434 Road Vehicles – Cybersecurity Engineering
- NIST 800-171 Security controls and processes for data protection
- Cybersecurity Maturity Model Certification (CMMC) program
- European Cyber Resilience Act and more to come

Auditing the compliance with the requirements defined in these documents requires subject matter expertise and can be time-consuming. Therefore, it is often outsourced. With more and more services in the cloud, there are also approaches to check the compliance with specific requirements fully automated [19]

[7] https://www.smh.com.au/technology/heartbleed-disclosure-timeline-who-knew-what-and-when-20140414- zqurk.html

[8] https://isc.sans.edu/diary/26798

9.3 Future of CSaaS

Future CSaaS offerings will potentially have to consider different currently ongoing trends in the security landscape. For example, there is the already mentioned threat of AI-enhanced malware. However, AI also poses other security threats to companies and public organizations, e.g., in the form of deep fakes or ChatGPT-generated spear phishing campaigns. Considering the ease of use of tools like ChatGPT, tailored spear phishing could have the potential to supersede normal spam not only in terms of quality but also in numbers. Another trend is the increasing number of supply chain attacks [20]. This may lead to an increased demand for zero trust architectures (ZTAs), especially toward previously trusted third parties, which can be potential starting points for the mentioned supply chain attacks, as well as for enhanced protection of customer data needed to deliver specific managed security services, e.g., SIEM. Moreover, existing trust relationships, for example, toward critical information infrastructures such as certificate authorities (CAs), have to be reconsidered, and enhanced control mechanisms need to be established [21]. Eventually, the rise of quantum computers may not directly lead to new types of services. However, it will definitely have an impact on existing services. They will have to timely adapt to the new post-quantum algorithms once they are finally standardized by NIST to ensure future-proof security is also protecting against store now, decrypt later type of threat scenarios.

9.4 Findings and Suggestions

The fact that 95% of all company malware breaches are caused by human error [9] has precipitated in the volume of companies currently adopting cyber-awareness training programs to increase by 15% year on year to date and the cyber-awareness training market to reach a predicted $10 billion annually by 2027 [22]. Additionally, the number of companies opting to pay for cyber-insurance has risen steadily over the last 3 years partly due to a large number of high profile attacks during this timeframe and the war in Ukraine. However, the uptake has now started to level off mainly due to the estimated 83% hike in cyber-insurance premiums over the last 12 months and the purchasing of better IT security equipment (e.g. next-gen firewalls and business continuity solutions) [23]. As working from home, either partly or totally, has become more mainstream for employees around the world, companies have had to look at new ways to protect their employees and intellectual property from malware attacks. As company IT staff cannot effectively protect all of these new remote working locations, decision-makers are opting for CSaaS companies to assist with this large threat canvas. This new working model bodes well for the future growth of the IT security services industry. Finally, the new elephant in the room, from a security threat perspective, is the mobile phone. These ultra-portable computers can now handle most of the day-to-day employee tasks like answering e-mail, attending meetings, and workflow approvals to reading and writing company documents. Most companies still overlook the security threat that mobile phones pose. They are finally taking action by installing anti-malware protection on these devices, allowing them access to guest wireless networks only and banning them from company meetings.

9.5 Conclusion

It is important to mention that the protection demand of a specific organization can be highly individual depending on factors, such as the sectors they are doing business in and the type of data they manage. The list of security services therefore only covers a selection of services which are most likely to be relevant for the majority of companies. When deciding which protection needs are applicable for an individual organization, it is recommended to include representatives of the organization's stakeholders and utilize independent advice from external specialists, where needed. Companies employing connected manufacturing processes in the context of Industry 4.0, for example, might have an increased demand for monitoring focusing particularly on industrial control system (ICS) or operational technology (OT) which implies factors like safety and therefore another kind of security goal prioritization. Explaining such sector-specific demands is not within the scope of this chapter.

In Table 9.1, the different services described throughout this chapter are mapped to the initially mentioned top ten companies in the managed and professional security services market in Europe according to Frost and Sullivan. It shows that almost all services are delivered by most of the discussed companies with just a few exceptions. One outstanding exception is cyber-insurance. That is because cyber-insurance is traditionally provided by traditional insurance companies rather than by tech companies specializing in cybersecurity services. However, representatives of both sectors do closely collaborate, e.g., regarding consulting and incident response services, as already described in the corresponding section of this chapter. There are even product bundles such as Deutsche Telekom's "Magenta Security Shield" which includes technical monitoring and response services as well as cyber-insurance. Although a bundled offer, the latter is, however, backed by the Allianz insurance company.

Table 9.1 is based on open-source intelligence, leveraging marketing channels such as the vendor's websites, service brochures, and white papers which are publicly available via the Internet. If vendors are not mapped to a specific service, it does not necessarily mean that they are not offering this service. Rather, it means that no information regarding this service from the specific vendor could be found at the point in time our investigation took place. Ultimately, what this all means is that the demand for CSaaS and additional security services will increase in tandem with the expanding threat landscape that has created a real sense of fear across the entire IoE landscape.

Table 9.1 Mapping CSaaS to top ten professional security providers according to Frost and Sullivan [7]

	Accenture	Capgemini	Orange Cyberdefense	Atos	IBM	Infosys	KPMG	Computacenter	Deutsche Telekom	Telefonica Tech
Security personnel as a service	○	○	●	●	●	○	●	○	◐	○
Cyber-awareness training	○	●	●	○	●	●	●	○	●	○
Vulnerability assessment	●	○	●	●	●	●	●	●	●	●
Periodic penetration testing	●	●	●	●	●	●	●	●	●	●
E-mail security	○	○	●	●	●	●	○	●	●	●
IAM	●	●	●	●	●	●	●	●	●	●
Cyber-insurance	○	○	○	○	○	○	○	○	◐	○
Incident response	●	●	●	●	●	●	●	●	●	●
BCDR	●	●	●	●	●	●	●	●	●	○
SIEM	●	●	●	●	●	●	●	●	●	●
System patching and updates	●	◐	○	◐	○	◐	○	●	○	●
Security standards compliance	●	●	●	●	●	●	●	●	○	●

References

1. Cloud Security Alliance – Security as a Service Working Group. Defined Categories of Security as a Service. https://downloads.cloudsecurityalliance.org/assets/research/security-as-a-service/csa-categories-securitiesprep.pdf. 2016
2. Sahay M (2023) Who Invented the Antivirus? A History of Antivirus Software. https://downloads.cloudsecurityalliance.org/assets/research/security-as-a-service/csa-categories-securities-prep.pdf, 2016
3. Statista Annual number of data compromises and individuals impacted in the United States from 2005 to 2022. https://www.statista.com/statistics/273550/data-breaches-recorded-in-the-united-states-by-number-of-breaches-and-records-exposed/. 2023
4. Johnson R (2019) percent of small companies close within 6 months of being hacked (2019). https://www.isc2.org//-/media/ISC2/Research/
5. Kirat D, Jang J, Stoecklin M (2018) Deeplocker–concealing targeted attacks with ai locksmithing. Blackhat USA 1:1-29
6. WorkForce-Study/ISC2-Cybersecurity-Workforce-Study.ashx (2022) International Information System Security Certification Consortium (ISC)².Cybersecurity
7. Sullivan F (2022) European Managed and Professional Security Services Market. https://store.frost.com/european-managed-and-professionalsecurity-services-market.html.
8. Statista. (2023) Size of the Security as a Service (SECaaS) market worldwide from 2022 to 2033. security-services-market. https://www.statista.com/statistics/595164/worldwidesecurity-as-a-service-market-size/.
9. Security I (2022) X-Force Threat Intelligence Index 2022. Page 16, Sum of top infection vectors linked to human error cas. https://www.ibm.com/downloads/cas/ADLMYLAZ
10. Security. H (2020) New Office 365 phishing tactics are difficult to spot but easy to prevent. https://www.rmtechteam.com/blog/new-office-365-phishingtactics-are-difficult-to-spot-but-easy-to-preve
11. Security H (2023) Next-Gen Vulnerability Management. https://www.holmsecurity.com/
12. Khan RA, Khan SU, Khan HU, Ilyas M (2021) Systematic mapping study on security approaches in secure software engineering. IEEE Access 9:19139–19160. https://doi.org/10.1109/ACCESS.2021.3052311
13. Hardt D (2012) The OAuth 2.0 Authorization Framework. RFC 6749. https://www.rfc-editor.org/info/rfc6749.
14. Sakimura N (2014) OpenID Connect Core 1.0. http://openid.net/specs/openid-connect-core-1%5C_0.html
15. Campbell B, Mortimore C, Jones M (2015) Security assertion markup language (SAML) 2.0 profile for OAuth 2.0 client authentication and authorization grants. https://www.rfc-editor.org/rfc/rfc7522
16. Mclean M (2023) Must-Know Cyber Attack Statistics and Trends. 2023. url:. https://www.embroker.com/blog/cyber-attack-statistics/.
17. Beamer T (2022) What Industries Are Most Vulnerable to Cyber Attacks In 2022? 2023. https://www.techbusinessnews.com.au/what-industries-are-mostvulnerable-to-cyberattacks-in-2022/
18. Cichonski P, Millar T, Grance T, Scarfone K (2012) Computer security incident handling guide. NIST Special Publication 800 (61):1–147
19. Stephanow P, Banse C (2017) Clouditor-continuous cloud assurance. Technical report, Fraunhofer AISEC. https://doi.org/10.1145/3338466.3358917
20. Hogben G, Dekker M (2012) European union agency for network and information security (enisa). Procure Secure-A guide to monitoring of security service levels in cloud contracts, Brüssel

21. Heinl MP, Giehl A, Wiedermann N, Plaga S, Kargl F MERCAT: A metric for the evaluation and reconsideration of certificate authority trustworthiness. In: Proceedings of the 2019 ACM SIGSAC Conference on Cloud Computing Security Workshop, 2019. pp 1–15
22. Morgan S (2023) Security Awareness Training Market To Hit $10 Billion Annually By 2027. https://cybersecurityventures.com/security-awarenesstraining-market-to-hit-10-billion-annually-by-2027/
23. Ankura (2023) The Cybersecurity Insurance Market: What to Expect in 2023. Mar. 2023. url:. https://www.jdsupra.com/legalnews%20/the-cybersecurity-insurance-market-what-2446460/.

Chapter 10
Big Data Analytics for Cybersecurity in IoE Networks

Faisal Rehman, Hanan Sharif, Muhammad Anwar, and Naveed Riaz

10.1 Introduction

Large-scale information processing and storage are two common applications of cloud computing [1]. Participant-to-participant, participant-to-machine, and individual-to-individual contacts are established as concepts like the Internet of Everything (IoE). For software analysis, it is crucial to merge the IoE systems and Big Datasets. Multiple sensors are utilized in IoE to collect data. Large-scale information analysis is used to collect, combine, and use data to enhance living standards [2, 3]. While carrying out this process, security and protection must be guaranteed.

Technologies like Big Data and IoE are at the height of exaggerated evaluation and are anticipated to continue deployed. IoE produces enormous amounts of heterogeneous digital information. Technologies for big information analysis may help with effective data management, retention, and evaluation. Big Data analytics must be used in conjunction with IoE since IoE alone cannot handle the full interaction

F. Rehman
Department of Computer Science and Information Technology, Lahore Leads University, Lahore, Pakistan

Department of Statistics and Data Science, University of Mianwali, Mianwali, Pakistan

H. Sharif
Department of Computer Science and Information Technology, Lahore Leads University, Lahore, Pakistan

M. Anwar (✉)
Department of Information Sciences, Division of Science and Technology, University of Education, Lahore, Pakistan
e-mail: anwar.muhammad@ue.edu.pk

N. Riaz
National University of Sciences & Technology, Islamabad, Pakistan

K. Naseer Qureshi et al. (eds.), *Cybersecurity Vigilance and Security Engineering of Internet of Everything*, Internet of Things,
https://doi.org/10.1007/978-3-031-45162-1_10

Fig. 10.1 Big Data and analytics advantages for different IoE applications

between its components and derive value from the information. IoE systems provide a developer-friendly user interface; are saleable, simple to use, and flexible for implementation; and have a clear system design. Figure 10.1 shows the Big Data and analytics advantages for different IoE applications.

10.2 Big Data Analytics

The term "Big Data" was coined as a solution to the rapid development of digital data [4]. The data is analyzed and used in a way that fosters creativity, increases productivity, and enhances decision-making. The term "Big Data" is used by Gartner to describe information assets with a high quantity, rate, and diverseness that require creative, efficient methods of information processing to improve apprehension and judgment calls. Large-scale data analytics has several benefits, including the capacity to create new goods and services, reduce costs, act quickly, and make better decisions. The vast amounts of data produced are susceptible to security flaws. It is crucial to protect this information. Data must be protected at all stages, including accumulation, transmission, and depository. The developments and difficulties in Big Data analytics and using the technology for safety were highlighted in [5]. The Big Datasets are a simple tool to use for credit card usage fraud detection. Issues with data management arise because of the constant growth in data size [6]. Data security is put at risk by data mining without approval or notice. Big Data is also accessible because of the platform's cluster-based feature. The fundamental elements of Big Data analytics are shown in Fig. 10.2.

Fig. 10.2 Big Data analytics

To solve these security problems, several technologies have been utilized and examined and are being developed. A few crucial techniques for safeguarding Big Datasets during analysis include encryption, logically centralized administration, access control for users, invasion detection, and security systems. HADOOP, MapReduce, HDFS, Hive, HCatalog, HBase, PIG, Mahout, Cassandra, In-Memory, NOSQL, and other analytical approaches are utilized with Big Data. Big Dataset architectures must meet several high-level criteria, including advanced analytic tools, versatile storage choices, and dependable data integration. The important aspects of deep learning include relevance, accuracy, timeliness, and management of data. Using methods like encryption, verification, tagging, labeled information, unorganized transmission, privacy protection, tracking, and assuring standards and legality, privacy protection may be offered in Big Data.

The IoE relies on linked information, processes, persons, and objects [7]. It has an impact on enterprises, industrial purposes, and people's daily lives. Real-world data collected from a variety of devices is connected and used in process automation with a focus on people [8]. IoE helps achieve social, financial, sustainable development and social policy goals. Additionally, it is utilized in several other sectors, including the extraction of fossil fuels, automation, e-learning, monitoring systems, smart grids, and traffic control and management [9]. In August 2018, Google reported 33.4 million entries for "IoE data sources" and over 346,000 entries for "IoE sources of data." Exabytes of data are generated daily from the IoE, according to "The Global Information Technology Review 2014: Benefits and Threats of Large Information." The IoE framework contains procedures, information, and people. As a result, every day, enormous amounts of information are generated. While conventional procedures can be utilized to handle data from the IoE, they might not be suitable for every type of data. As a result, research is being conducted to ensure the security of the information generated by IoE.

10.3 Securing IoE with Big Data Analytics

By 2020, there may be 100 billion linked devices, according to the IoE. The IoE will produce a lot of data. With regard to ICT infrastructure, such data may be connected to several problems and difficulties. With regard to ICT infrastructure, such data may be connected to several problems and difficulties. To handle and utilize this generated data, Big Data analytics is required [10]. "Big Data" is sometimes referred to as data gathered about the truth. These facts are produced by sensors that are inserted into the objects that encircle us. The more information that is accumulated, the more it may be exploited to advance technology. That's why collecting information and tracking are ongoing processes; it is crucial to maintain these instruments operational and continually linked to the Internet to guarantee continuous data upgrades to the web server. As a result, we are more susceptible to intrusions that might compromise the information's privacy and security.

Velocity, density, and diversity are three terms that are used in [11] to describe large amounts of information. Other important characteristics of Big Datasets include variability, truthfulness, and value of the data. It indicates the quantity of data produced, the rate at which it is produced, and the fact that it is accessible in many formats. This huge amount of data is kept on the cloud. Some businesses utilize this information to research users' browsing and purchasing patterns. For many users, this can raise serious privacy issues. The data is vulnerable to security breaches and data leaks. If these data breaches are not appropriately handled, they harm the image of major organizations. The IoE uses technologies that are not designed for secure communication. As a result, both the network and the data are exposed. In many instances, Big Data analytics deals with this problem.

Large amounts of information that is correct and produced by IoE. The IoE ecosystem may even have data altered by cybercriminal attackers, causing instability and unrest. The production of Big Data by the IoE is represented by the rising volume of enormous amounts of diverse data. With the aid of safety and a conceptual architecture for information gathering, transportation, and retention, [12] suggest merging IoT, big information analysis, and complicated event processing methods to address the key data management challenges in the healthcare sector. They proposed a fully functional, integrative medical system. Big Dataset analysis and complicated event processing methods are used to address the key data management challenges in the healthcare sector. They proposed a fully functional, integrative medical system. Large-scale information analysis and complicated processing techniques are used to address the key data management challenges in the medical industry. They proposed a fully functional, integrated medical system.

10.4 Related Work

Recent years have seen the development of several safe Big Data solutions. These solutions are shown in Table 10.1, which also highlights the solution's breadth, methodology, and complementary techniques (such as statistics or machine learning) that it supports.

The answer discovered by employing Hadoop and Apache Spark, as well as those who have thought about enhancing the usage of large amounts of information with other solutions like data analysis or deep learning, is that Hadoop and Apache are the large information datasets most frequently used for various scientific ideas. Recent years have seen the development of several large-scale cybersecurity technologies. These solutions are shown in Table 10.2, which also highlights the solutions' breadth, innovation, and complementary methodologies (such as statistics or machine learning) that they support. Additionally, the cybersecurity activities are carried out utilizing information systems, including AD, NA, AC, ID, CTI, and ATD. The suggestions that have been examined and the security events have the same range, much like monitoring tools and networking. ATD groups suggest for DDoS and phishing detection. As can be seen, alert correlation and cyber-threat intelligence are less formulated, whereas attack perception is the key focus of most cybersecurity operating apps.

Table 10.1 Big Data proposal

ID	Scope	Technology	Complement	Author
S1	Anomaly detection	Apache spark	Social media	[12]
S2	Network monitoring	Hadoop	None	[13]
S3	Network monitoring	Hadoop	None	[14]
S4	Instruction detection	Hadoop	GPGNU	[8]
S5	Anomaly detection	Apache spark	Machine learning	[15]
S6	Anomaly detection	Apache spark	Machine learning	[16]
S7	Intrusion detection	Apache spark	None	[17]
S8	Anomaly detection	Hadoop	None	[18]

Table 10.2 Apache Metron modules

Module	Solution
Data access	Hive Solr Hbase
Steam process and enrichment	Spark storm
Message queue	Kafka
Data collection	Pcap

10.4.1 Big Data Commercial Solutions for Cybersecurity

WCS [19] combined two of its trademarks: QRadar Advisor, an event and security information management tool, and Watson, a learning engine that employs human language technology to evaluate ambiguous data, such as Internet site information. QRadar combines the activities from several data sources, including machines, web servers, and gateways. By utilizing Watson, it is possible to compare local security information from QRadar with unstructured information from blogs, sites, and scientific literature. A real-time surveillance platform is made up of three major parts: an actual computing processor, telematics information sources, and satellite tracking data collectors. Apache Metron is the latter.

The real-time system CDH is based on Apache Hadoop. A software platform called Apache Hadoop enables distributed applications to run across multiple computers and analyze enormous data volumes using straightforward design patterns. Downloading stream sets, configuring data pipelines, and setting Apache Spots ODM in HDFS are the three macro phases that make up the cluster configuration. Based on the Apache Spot Open Data Structure, CDH for managing data considers the Server Reference Manual, Security Assessments, Windows Firewall Records, and Tertiary Circular Authentication Gateway Computer Records as data sources. Six core data categories are specified by the CDH architectural style.

SELKS is an open Linux operating system that employs the Elasticsearch stack to connect and show security alerts and is built on the code ecosystem for the detection of invasions. The SELKS's constituent parts are the following:

- Suricata is a higher-bandwidth IDS that can handle data rates of up to 10 GB/s.
- Logstash analyzes the various sources of data.
- Elasticsearch handles the scanning of data sources.
- An elastic search element can be used to read data from Kibana, a visualization tool that lets users build bespoke displays.
- Kibana may use Scirius, a Suricata web interface, to draw patterns from Scirus.

Table 10.3 summarizes the characteristics of each solution that we believe are most important, including RTP, NLP, IDS, ML, VA, CD, and ES (such as blogging and web sites).

10.5 Processing Methodology Using Big Data

When processing large amounts of data, a comprehensive approach that starts with the business issue and ends with the usefulness of the analytical model should be taken into account. A data processing model typically consists of several stages, including data acquisition and licensing (information knowledge), excavation, washing, and information, assimilation, agglomeration, and depiction (treatment), simulation, visual analytics and explanation, interaction, application, and judgment call. Several elements make up the Big Dataset processing approach, which enables

Table 10.3 Important characteristics of Big Data cybersecurity solutions

Attribute	Watson	Hortonworks	Cloudera	Selks
CD	No	No	No	Yes
Ml	No	No	No	No
ES	Yes	Yes	No	No
NLP	Yes	Yes	No	No
Open	No	Yes	Yes	Yes
RTP	Yes	Yes	Yes	Yes
IDS	Yes	No	Yes	Yes
VA	Yes	No	Yes	No
Core	Watson	Spark	Hadoop	ELK

the conversion of information into understanding. Data mining is regarded as the most significant stage of the KDD procedure because it brings together the methods for analyzing the data that is already accessible. Knowledge can be derived from the usage or comprehension of the created model. An "underground mining perspective" of the data, which can be seen as a transition within the designed system in Sect. V, is necessary to use methods of data mining [20] (designs). This view includes several stages where we find the assessment of the probability density of each attribute in systems designed to check values.

Variable transformation should be done, and it needs to be stated that whether the values are disregarded, discarded, or modified will depend on the issue to be addressed and the information analysis strategy to be used. Procedures must be taken to attain vitality [21].

10.6 Cybersecurity Architecture Based on Big Data

The subjects addressed in this area are those where Big Datasets may benefit security. Then, a five-layered structure as presented in [22] consists of an extracting layer, a loading layer, a conversion layer, and an implementation layer. This structure makes the claim that it can spot unusual activity trends and patterns to predict cybercrime attacks that are described as being spontaneous, random, and unusual. This chapter indicates that huge amounts of information mostly concentrate on examining changes and assaults, but they are passive information security measures with the goal of producing warnings for the security professional. The implementation of active security measures like computer security and danger hunt makes use of Big Data analysis to forecast potential assaults in the future. This allows assault tendencies to be identified and hackers to be identified, allowing for the creation of response plans. Big Data helps in analyzing both organized and unstructured information, including records, images, and movies that are utilized as forensic data in cyber-forensics procedures. An example of the subjects where Big Dataset mining might benefit cybersecurity [23] is shown in Fig. 10.3.

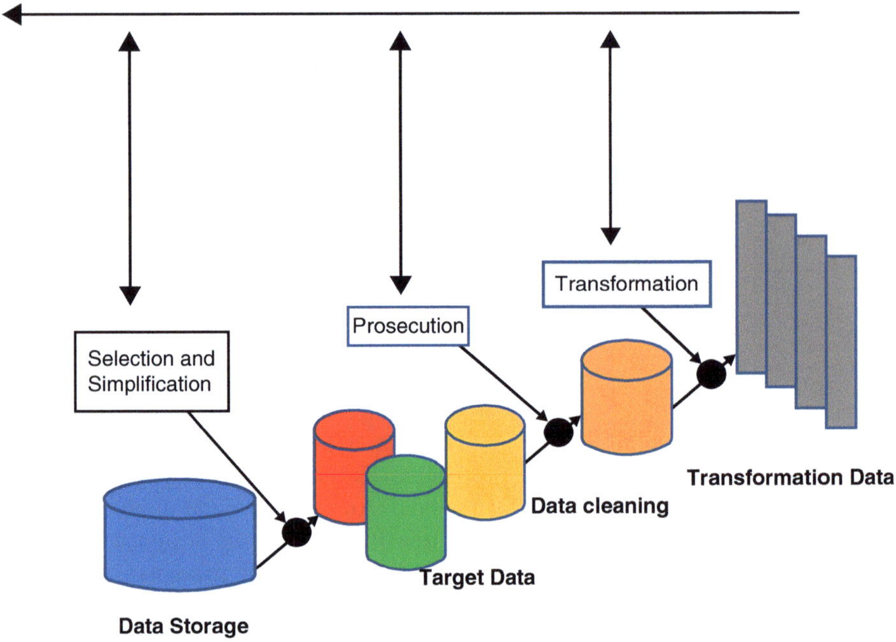

Fig. 10.3 Mineable view

The protection, examination, and evaluation of digital evidence are the main goals of forensic examination. In 2016, 17,088 pieces of proof were received, according to the 2016 Central Digital Forensics Lab report from the FBI. A total of 5667 gigabytes were produced because of forensic digital analysis. According to the researchers' definition, Big Data analytics is a subset of computer evidence that deals with huge databases during the detection, gathering, classification, and presentation stages. Additionally, they offer a theoretical design for Hadoop-based investigations that considers a resilience layer to eliminate data duplication. In Big Data ideas, this is a severe issue for guaranteeing data quality and authenticity and preventing inaccurate outcomes because of duplicate content.

The authors state that using visual methods can shorten the search time and increase the efficiency of finding questionable files. In the modern Internet era, an analyst must examine vast amounts of data from numerous sources. Big Data solutions offer two important ways to use data: (i) combining data from many sources with different file types, like photos, words, or movies, and (ii) making custom graphical representations with geographical coordinates that make it easier for researchers to access more important information. For malware recognition, the Internet of Things targeted 120,000 different malware variants in the first half of 2018 [24]; thus it is important to introduce new technical advancements considering the expansion of data and the requirement to speed up computing. This situation piqued the interest of several researchers in investigating the use of Big Datasets for detecting attacks. The authors of [25] present a modular grouping method that uses more than 75,000 samples and takes 3 h to process to find and classify malware with similar behavior. The authors present a technique for categorizing malware by combining Big Data analysis, binaries monitoring, and dynamical command-flow analysis. The authors discuss concerns and difficulties with malware detection, including adversarial learning, active learning, malware prediction, ratio, and incremental learning. Security breach comprises two basic methods: threat hunting and cyber-dissimulation.

The goal of cyber-deception is to identify attacks so that appropriate cyberspace defensive strategies can be developed to fool the hackers. Honeypots and honeynets are common cyber-deception tools, but some intriguing research goals in this area include using AI, game theory, and Big Datasets to improve security defenses against hackers. Rather than wait for attack alerts, active defense employs threat hunting, an initial phase that searches through networking and security metadata to find threats. The most significant contributions from these two researches may be connected and also inferred that threat search is concentrated on identified and unidentified dangers. Before launching an assault, weaknesses and attack procedures are identified utilizing fundamental search methods, data methods, visual analytic techniques, collection, and Bayesian probabilities [26].

The amount of information that must be processed throughout the anti-malware process is greater than what a normal being can handle. It is easy to overcome this limitation by adopting Big Data technologies. For attack recognition, to decrease the time between attack discovery and response, intelligence experts must identify assaults as quickly as possible. An acceptable false-positive rate is necessary for successfully detecting attacks. PCA and MDRA are the writers' two suggested detection strategies [27]. The writers suggest Apache Spark-based automated detection methods utilizing PCA for magnitude reduction. They point out that choosing a collection of skills, scaling, and evaluating the gained information are difficulties faced by data applications.

10.7 Data Analytics Architecture for Cybersecurity Applications

The extracting layer, the loading layer, the conversion layer, and the implementation layer are the five functional layers that make up the suggested design. Cyber-security attacks are described as being largely random, unplanned, and unusual in nature. The various layers are merged to uncover abnormal behavioral patterns. Figure 10.4 shows the architecture for Big Data security.

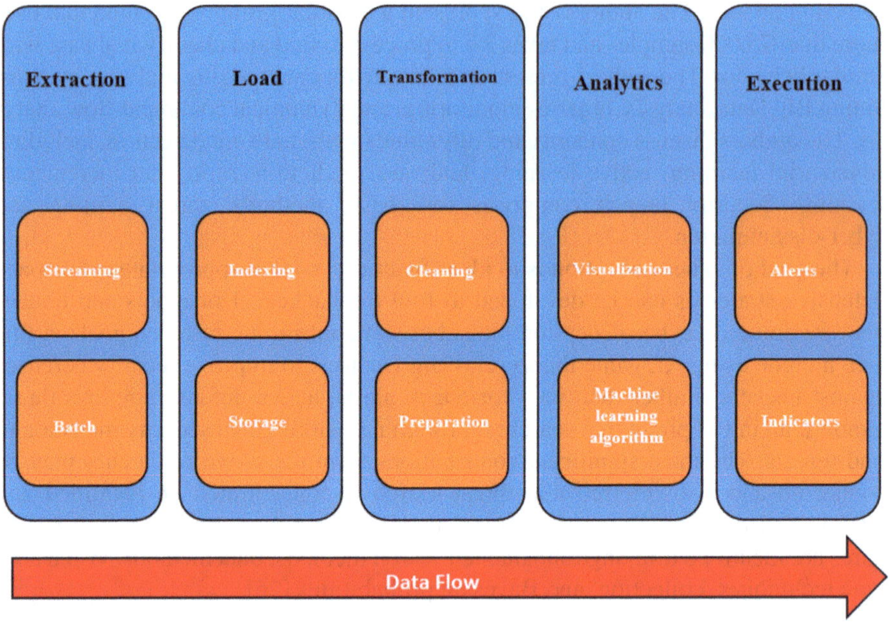

Fig. 10.4 Architecture for Big Data security

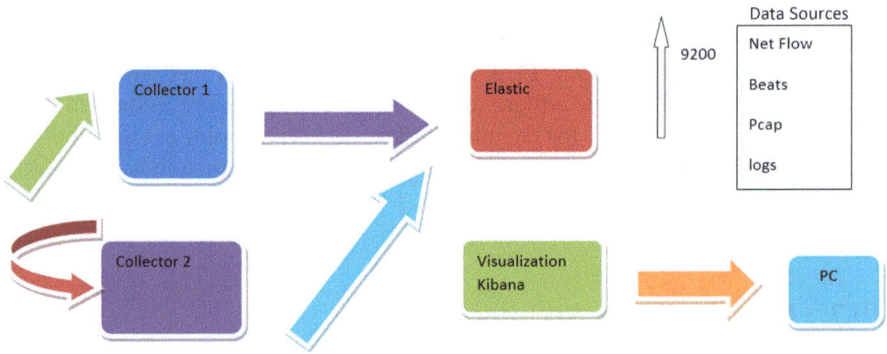

Fig. 10.5 Implementation of the architecture

10.7.1 Indicators Module

The indications component will enable the display of important performance metrics so that the state of the data can be determined very quickly. The architectural design using the ELK stack to cover the collection, loading, and implementation layers is shown in Fig. 10.5. The quantity of datasets affects the number of collected servers. The batching process can reach its limit because of the huge amount of information in information security. The broadcast method works well for data extraction. Data processing for collection servers happens instantly. Log, NETCONF, or rhythms are a few examples of security datasets that may be used to identify unusual activity patterns. Data and signs of compromise are the two key components that determine the visualization layer's IOC. The first one is connected to the kind of data that could generate details about bizarre behavior. There are many distinct information sources, including firewalls, networks, servers, and terminals. Attempt to analyze all this data to boost the capacity of the Big Data design overall and the number of collection devices.

The second is based on the very first: it is impossible to produce helpful warnings concerning disorder if the data derived from data sources is not useful. The IoC enables the security researcher to determine whether a certain occurrence is harmful or safe. The number of connections that were utilized over a certain period, and the safety specialist was unable to quickly determine whether this is a sign of a cyber-assault [28]. This number was produced using NetFlow traffic. DNS traffic is still another scenario. This sort of data can be processed using the ELK structure. Without sufficient IoCs, a cybersecurity investigator cannot determine whether large connectivity exists. In this case, the researcher must assess the history to determine whether the current DNS makes a good count. This may be an important consideration in the broadcasting procedure when computational power is more important than information storage. Machine learning techniques might be thought of as a replacement to fill this gap. In the ELK structure, data might be combined. As an illustration, domain data and NetFlow data may be connected, enabling a cybersecurity researcher to link DNS queries to the Internet protocol. This factor can be important to pinpoint the attacker's geographic location.

10.8 Discussion

The complicated and dynamic surroundings resulting from technological and social development generate huge amounts of datasets, which presents new problems for cybersecurity experts that evaluate this information to find trends or irregularities that allow for detecting danger or cyber-assaults. Large-scale data analytics is suggested as a fresh approach to boosting the efficiency of security services by quickly evaluating massive amounts of data in various forms. A large amount of data is mostly used in cybersecurity to analyze activities and spot irregularities while concentrating on defensive security measures. Large-scale data analysis may improve other security operations for preventive measures like threat tracking or digital fraud. Big Data may be used in conjunction with other approaches to improve its ability to analyze massive amounts of information from various sources to identify the attack. For example, through analyst training, machine learning makes it possible to automate the process of spotting abnormalities, while natural language processing (NLP) makes it possible to find identifying patterns in blog posts or news articles from security news sites.

Keep in mind that the ELK might analyze a variety of sources of data. The data must first be cleaned. Since the ELK structure does not have a method to analyze this sort of data in its basic setup, encrypted traffic is another important factor to think about. It is critical to specify the issue that must be resolved or that is antagonistic to the design, as this will necessitate the use of specific data sources and variables. It is important to understand the economic aspect since it enables resolving the issue at hand. It is also advised to collaborate with corporate entities during this phase. The architecture that enables processing massive volumes of data, like communication systems or server logs, must keep leveling and fast-read drives in mind.

10.9 Conclusion

The development of a safe IoE architecture based on information analysis is thoroughly evaluated in this section. Additionally, it gives a snapshot of the huge data production from IoE-related gadgets, people, and sensing elements. The idea of a linked and intelligent community is presented. These technologies are used in urban, medical, factory equipment, and other various domains, along with the research that has been done in these areas. By separating the functions of Big Data into various platforms, such as digital networking, business analysis, cloud computing, and others, future expansion may be achieved. Another key element to think about is data security while transferring from IoE to the Internet or other devices.

References

1. Dias I, Sousa MJ Business Intelligence applied to Human resources management. In: New Contributions in Information Systems and Technologies: Volume 2, 2015. Springer, pp. 105–113
2. Rathore MM, Ahmad A, Paul A, Rho S (2016) Urban planning and building smart cities based on the internet of things using big data analytics. Computer networks 101:63–80. doi:https://doi.org/10.1016/j.comnet.2015.12.023
3. Kashif Naseer Qureshi AA, Raja Waseem Anwar, Shahid Nazir Bhati, and Gwanggil Jeon (2021) Fully Integrated Data Communication Framework by Using Visualization Augmented Reality for Internet of Things Networks. Big Data 9 (4):253–264. doi:https://doi.org/10.1089/big.2020.0282
4. Zhan K (2021) Sports and health big data system based on 5G network and Internet of Things system. Microprocessors and Microsystems 80:103363
5. Pramanik M, Lau R, Youe W, Ye Y, Li C (2017) Big data analytics for security and criminal investigation. WIREs Data Min Knowl Discov 7: e1208. doi:https://doi.org/10.1002/widm.1208
6. Qureshi MA, Qureshi KN, Jeon G, Piccialli F (2021) Deep learning-based ambient assisted living for self-management of cardiovascular conditions. Neural Computing and Applications. doi:https://doi.org/10.1007/s00521-020-05678-w

7. Shojafar M, Sookhak M (2020) Internet of everything, networks, applications, and computing systems (IoENACS). vol 42. Taylor & Francis. doi:https://doi.org/10.1080/1206212X.2019.1575621
8. Bandre SR, Nandimath JN Design consideration of Network Intrusion detection system using Hadoop and GPGPU. In: 2015 international conference on pervasive computing (ICPC), 2015. IEEE, pp 1–6. doi:https://doi.org/10.1109/PERVASIVE.2015.7087201
9. Qureshi KN, Qayyum S, Ul Islam MN, Jeon G (2021) A secure data parallel processing based embedded system for internet of things computer vision using field programmable gate array devices. International Journal of Circuit Theory and Applications n/a (n/a). doi:https://doi.org/10.1002/cta.2964
10. Ahmed E, Yaqoob I, Hashem IAT, Khan I, Ahmed AIA, Imran M, Vasilakos AV (2017) The role of big data analytics in Internet of Things. Computer Networks 129:459–471. doi:https://doi.org/10.1016/j.comnet.2017.06.013
11. Lee I (2017) Big data: Dimensions, evolution, impacts, and challenges. Business horizons 60 (3):293–303. doi:https://doi.org/10.1016/j.bushor.2017.01.004
12. Sheriff CI, Naqishbandi T, Geetha A Healthcare informatics and analytics framework. In: 2015 International Conference on Computer Communication and Informatics (ICCCI), 2015. IEEE, pp 1–6. doi:https://doi.org/10.1109/ICCCI.2015.7218108
13. Marchal S, Jiang X, State R, Engel T A big data architecture for large scale security monitoring. In: 2014 IEEE International Congress on Big Data, 2014. IEEE, pp 56–63. doi:https://doi.org/10.1109/BigData.Congress.2014.18
14. Chen Z, Zhang H, Hatcher WG, Nguyen J, Yu W A streaming-based network monitoring and threat detection system. In: 2016 IEEE 14th International Conference on Software Engineering Research, Management and Applications (SERA), 2016. IEEE, pp 31–37. doi:https://doi.org/10.1109/SERA.2016.7516125
15. Lighari SN Testing of algorithms for anomaly detection in Big data using Apache spark. In: 2017 9th International Conference on Computational Intelligence and Communication Networks (CICN), 2017. IEEE, pp 97–100. doi:https://doi.org/10.1109/CICN.2017.8319364
16. Jia Z, Shen C, Yi X, Chen Y, Yu T, Guan X Big-data analysis of multi-source logs for anomaly detection on network-based system. In: 2017 13th IEEE conference on automation science and engineering (CASE), 2017. IEEE, pp 1136–1141. doi:https://doi.org/10.1109/COASE.2017.8256257
17. Gupta GP, Kulariya M (2016) A framework for fast and efficient cyber security network intrusion detection using Apache spark. Procedia Computer Science 93:824–831. doi:https://doi.org/10.1016/j.procs.2016.07.238.
18. Fontugne R, Mazel J, Fukuda K Hashdoop: A MapReduce framework for network anomaly detection. In: 2014 IEEE conference on computer communications workshops (INFOCOM WKSHPS), 2014. IEEE, pp 494–499. doi:https://doi.org/10.1109/INFCOMW.2014.6849281
19. Dahiya P, Srivastava DK (2018) Network intrusion detection in big dataset using spark. Procedia computer science 132:253–262. doi:https://doi.org/10.1016/j.procs.2018.05.169
20. Vangara RVB, Vangara SP, Thirupathur V (2020) A survey on natural language processing in context with machine learning. Int J Anal Exp Modal Anal: 1390–1395
21. Fayyad U Data mining and knowledge discovery in databases: implications for scientific databases. In: Proceedings. Ninth International Conference on Scientific and Statistical Database Management (Cat. No. 97TB100150), 1997. IEEE, pp. 2–11. doi:https://doi.org/10.1109/ssdm.1997.621141
22. Srivastava J, Cooley R, Deshpande M, Tan P-N (2000) Web usage mining: Discovery and applications of usage patterns from web data. Acm Sigkdd Explorations Newsletter 1 (2):12–23. doi:https://doi.org/10.1145/846183.846188.
23. Rabhi L, Falih N, Afraites A, Bouikhalene B (2019) Big data approach and its applications in various fields. Procedia Computer Science 155:599–605. doi:https://doi.org/10.1016/j.procs.2019.08.084

24. Tien C-W, Chen S-W, Ban T, Kuo S-Y (2020) Machine learning framework to analyze IOT malware using ELF and opcode features. Digital Threats: Research and Practice 1 (1):1–19. doi:https://doi.org/10.1145/3378448
25. Ming J, Xin Z (2017) Impeding behavior-based malware analysis via replacement attacks to malware specifications. Journal of Computer Virology and Hacking Techniques 13 (3):193–207. doi:https://doi.org/10.1007/s11416-016-0281-3
26. Manogaran G, Thota C, Kumar MV (2016) MetaCloudDataStorage architecture for big data security in cloud computing. Procedia Computer Science 87:128–133. doi:https://doi.org/10.1016/j.procs.2016.05.138
27. Cheng J, Xu R, Tang X, Sheng VS, Cai C (2018) An abnormal network flow feature sequence prediction approach for DDoS attacks detection in big data environment. Comput Mater Continua 55 (1):95–119. doi:https://doi.org/10.3970/cmc.2018.055.095
28. Valle RR, Manzanares FV (2013) Concentration of goods traffic in Spanish ports during the period 2000–2009. Regional and Sectoral Economic Studies 13 (2):59–72

Chapter 11
Cybersecurity Standards and Policies for CPS in IoE

Kashif Naseer Qureshi, Garret O'Keeffe, Shane O'Farrell, and Graham Costelloe

11.1 Overview

Cybersecurity standards and policies are significant as a guideline and basic framework to protect the systems, networks, and other data processing components. Internet of Everything (IoE) is one of the new concepts where people, devices and processes, and systems are interconnected for data communication. These networks are further connected with backbone wired and wireless networks to collaborate in real time. The cybersecurity standards and frameworks can help to ensure the security and privacy of users and mitigate the potential risks and systems vulnerabilities. This chapter discusses the existing standards and frameworks to cover all Cyber-Physical Systems (CPS) for IoE networks. The chapter also suggests a standard framework to adopt and ensure confidentiality, integrity, and availability. The technical comparison of existing standards also discusses understanding the overall elements.

K. N. Qureshi (✉) · G. O'Keeffe
Department of Electronic & Computer Engineering, University of Limerick (UL), Limerick, Ireland
e-mail: kashifnaseer.qureshi@ul.ie; garret.okeeffe@mastercard.com

S. O'Farrell · G. Costelloe
Munster Technological University, Cork, Ireland
e-mail: shaneofarerell@mtu.ie; grahamcostelloe@mtu.ie

K. Naseer Qureshi et al. (eds.), *Cybersecurity Vigilance and Security Engineering of Internet of Everything*, Internet of Things,
https://doi.org/10.1007/978-3-031-45162-1_11

11.2 Introduction

A standard is an agreed way to build something, manage a process, or deliver a service for better processes and quality. Standards are represented as documents that define specifications, procedures, and guidelines, aiming to ensure the safety, consistency, and reliability of products, services, and systems. They are aggregated and distilled knowledge of the subject matter experts in the field who know the needs of the stakeholders they represent. Cybersecurity standards are designed to improve the security of IT systems, the networks they run on, and the infrastructure it is stored and processed on. Cybersecurity standards define the functional requirements to implement information security as well as the assurance requirements within the technology [1]. Cybersecurity standards are developed by cybersecurity subject matter experts to help people develop a system or assess an off-the-shelf or bespoke system to design or validate the application's security features [2].

As people, devices, and processes are involved in IoE networks and need proper security standards and frameworks to protect the user's data, standards need to be largely technology agnostic but must provide enough guidance to ensure the IoE system is as secure as possible without impeding the functionality of the system from doing its job. Standards cover a diverse set of areas, especially for IoE networks, and can range from a technical standard defining the cryptographic specifications for a crypto module to defining a process that ensures software is built in the recommended way (reducing the number of potential security flaws in the implementation) [3].

Both standards and guidelines provide guidance aimed at enhancing cybersecurity, but guidelines usually lack the level of consensus and formality associated with standards. Standards are a set of specifications that an organization should implement designed to reduce the risk to its clients. By implementing the standards, the company can categorically state that they have reached the quality as set out in the standard. In the case of a cybersecurity or data protection standard or regulation, this means clients of that organization can then be assured that their data is at least in some part secure against exfiltration, change, or misuse [4].

Standards provide a set of techniques, controls, and processes that they can implement to achieve and maintain a certain level of security. Standards also allow the organization to assess itself against a certain bar. Aligning with standards also helps a company when defining their approach to cybersecurity for themselves as they will have to build processes and mitigating controls specific to their organization to meet the standards they are trying to achieve. Standards tend to be created for organizations in specific industries and are used as a way of a) achieving a certain level of quality and b) assuring other clients or partners that they have met the level of quality needed to be trusted [5]. In the IoE networks, healthcare, education, transportation, and industrial companies are involved and need cybersecurity standards and frameworks to protect and secure systems. For example, the Institute of Electrical and Electronics Engineers Standards Association (IEEE-SA) develops standards in many areas, including information technology, telecommunications, and power generation. An example of IEEE-USA's security work is its 802 Local Area Network (LAN)/Metropolitan Area Network (MAN) Standards Committee [6].

The International Organization for Standardization (ISO) is a nongovernmental organization that comprises standards bodies from more than 160 countries, with 1 standards body representing each member country [7]. For example, the American National Standards Institute represents the United States. ISO members are national standards organizations that collaborate in the development and promotion of international standards for technology, scientific testing processes, working conditions, societal issues, and more. ISO and its members then sell documents detailing these standards [3]. There are many international, regional, national, industry, and government groups involved in the development of cybersecurity standards. Standards Developing Organization (SDO) is an organization whose primary mission is the development of voluntary consensus standards on an international, regional, or national basis. Most SDOs cover a wide variety of technical areas, not just cybersecurity. In many cases, several stakeholders from within an industry will come together to ally with the specific goal of writing a standard. An example of this is PCI-DSS which is a standard focused on improving payment account security by ensuring that all companies that accept, process, store, or transmit payment card data abide by it. This standard was brought about by an alliance of Visa, MasterCard, AmEx, Discover, and JCB [8].

Standards differ in the ways that they are regulated. Depending on which governing body or regulatory organization, compliance with standards may be optional, or compliance may be a requirement. Voluntary standards are generally called voluntary because their use is optional, although a regulating agency could adopt or mandate their use. Mandatory standards are standards whose use is prescribed by a regulatory agency or implementing organization. Mandatory standards typically implement laws and regulations [9]. For example, PCI is a mandatory standard for the payment card industry. Companies rarely only use one set of standards. Business problems are very often solved using a combination of technology, management, and business processes, and because of this, several standards will normally come into play to ensure the successful and safe implementation of the project. An example might be the PCI standards imposed by an IoE company developing a software product for credit card transactions, but the standards used for the network communications are also in play as well as those used for developing an overall information systems management strategy for the wider organization such as ISO27000 series [10].

An Information Security Policy should offer a framework from which an organization can implement all security controls and processes deemed necessary and enforceable. A framework is a bunch of tools, guidance, and resources to help an organization with how it should think about a certain goal and how to achieve it. A standard is much more specific in its criteria for achieving that standard (although often not giving guidance on how to achieve it). The IoE networks need standards and frameworks to ensure data security from external or internal sources. Figure 11.1 shows the IoE standards overview.

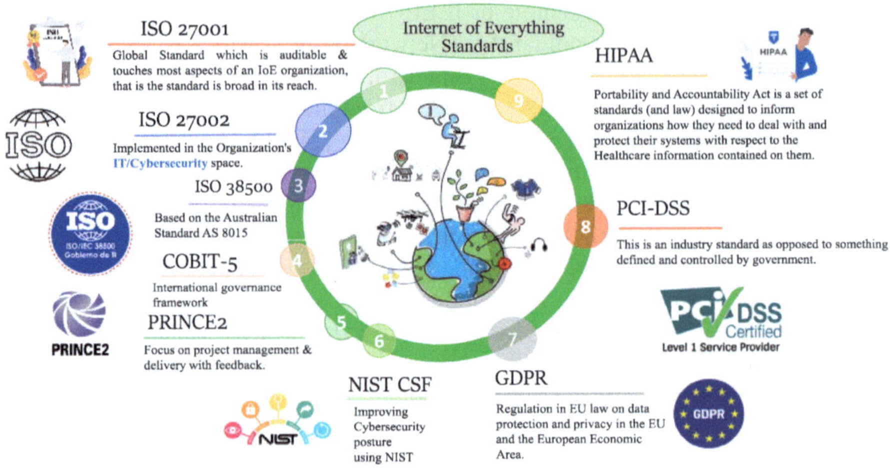

Fig 11.1 IoE standards overview

11.3 Information Security Standards Requirements, Policy, and Elements

Confidentiality, Integrity, and Availability (CIA) is the main principle of information security. Information security requirements for IoE networks should cover the following main area:

- Ensure user security by applying authorization and authentication to avoid unauthorized access to sensitive data.
- Ensure business continuity in any situation the business should run and normal.
- Timely identify the information security risks and come up with the risk management plan.
- Conduct training programs to make information security awareness to the organization.
- Ensure the data protection in IoE networks.
- Identify the new technology to protect the IoE systems.
- Identify and follow the industry standard to protect the data and organizations.
- It should include end-to-end security processes throughout the organization.
- The policy should be easy to understand and implement in heterogeneous IoE networks.
- Policy should be revised in a regular interval.
- Policy should focus on the organization goals.

11.3.1 Information Security Policy Elements

Purpose It covers overall approach of information security. This policy is for proper controls in IT department of an assurance company to ensure changes to production systems meet security standard and proper controls on production systems. The minimum number of people has access to production systems and data, ensuring data confidential for customers. The policy should set out the clear objectives for the information security. It should be able to set out how it will allow an organization to protect its IT or IoE networks assets, retain data integrity, be able to identify misuse of IT property (networks, assets), and protect it from security threats.

Audience It define the audience to whom the information security policy applies. The audience for this is the IT development team and production support team. This is not for system users. The policy should be able to identify the key stakeholders of the policy and also identify any high priority users whose policies might be more applicable depending on the data/responsibilities they may be working on/with.

Information Security Objectives Offer a secure, safe, data consistent environment and secure IoE systems from data breaches/threats by implementing a policy.

Authority and Access Control It defines hierarchical pattern and network security policy. The development team should only have access to development and test environments. Production support teams have access to production systems. Each user will have their own unique account, making their accounts individually traceable. By identifying the common users and the important users who will be working with more sensitive data and being able to authorize these users based on their relevant permissions or role-based access controls. There will be the physical control policies where certain uses will only be able to access certain physical areas of the organization.

Data Classification It classifies the data into categories "top secret," "secret," "confidential," and "public." Production environment is made up of multiple systems containing their own data. This data is to be reviewed for classifications such as health information only accessible to privileged users.

Data Support and Operations All confidential data must be encrypted at rest/storage and in transit. Backup of data is to be encrypted and stored in a secure location, with access limited to the backup team. All data transfer of confidential information must be encrypted and sent over a TLS connection. All data should be tagged with the relevant labels that will then associate their level of risk and only allow the specific users such as public, restrictive, confidential, and highly confidential data classifications. Numerous tools can be used to provide this element.

Security Awareness and Behavior Training is to be provided to the IT development team that how to develop best practices in the organization. Production support team is to be trained to handle confidential information. Awareness relies upon the sharing of knowledge with all staff not just the IT department or staff. Even having all the next-gen firewalls and security policies in place if a user clicks on a suspicious link and doesn't inform anybody, there is a massive threat of data breaches or an attack unfolding on the IT systems. The use of proxies can also help a business through web filtering and enforcing an acceptable Internet usage policy.

Encryption Policy Hide the data from unauthorized access. All disks containing confidential information must be encrypted.

Data Backup Policy Protect the data by making a copy of sensitive data in a secure environment. A full backup should to be taken every night and incremental every 10 min. Full backup should be stored in secure location limited to backup team and kept for 365 days.

Responsibilities, Rights, and Duties of Personnel It defines the responsibilities clearly. System to provide a report on the information held by us for an individual. System to provide the functionality to delete individuals from production systems. Helps to provide oversight on an organization's standard (ISO27001, ISO27002, COSO, CIS, and GDPR). There will be many different responsibilities which an organization will need to comply with. GDPR is the protection of personal data and the privacy of EU citizens. The security policy is responsible for protecting an organization's IT infrastructure.

11.4 Existing IoE Security Standards

This section discusses the most common existing cybersecurity standards and framework designed for communication systems and networks. These standards are also used for IoE networks because these networks are heterogeneous in nature and connected with backbone, clouds, and edge computing. These all systems are handled by organizations and companies.

11.4.1 ISO 27KX – ISO

This standard is the most commonly used set of standards in cybersecurity. These standards are generally concerned with the implementation of a certified information security management system within an organization. This means that the organization is doing its best and following best practices to ensure they are protecting user's data.

11.4.2 ISO 27001

This standard has the specifications for creating, operating, and controlling an ISMS. ISO 27002 then lists a structured set of controls to comply with 27001. This includes managing assets in an organization, securing human resources, managing operations and communications, securing environmental and physical aspects, managing business continuity, and managing compliance and information security incident areas [11]. The ISO standards also provide standards and guidance. ISO 27001 is an international standard that lays out a specification for an Information Security Management System (ISMS). This standard aims to address data security by focusing on people and processes and also technology same as in IoE networks. The standard has a heavy focus on its risk-assessment approach which stipulates that a risk assessment must be carried out before any controls can be selected and implemented. This standard follows a Plan-Do-Check-Act model and has an independently accredited certification to align the ISMS with information security best practices. ISO 27001 has an international presence that many organizations recognize and trust. The ISO 27001 primary focus is on information security controls, unlike COBIT which is considerably broader in scope, focusing on information technology governance. The primary benefits of implementing ISO 27001 are the following:

1. The identification of critical information through the detailed analysis.
2. The implementation of security controls following the analysis.
3. A completed information security risk assessment of the system under review.
4. These benefits all lead to developing and supporting a more secure culture in the organization.

11.4.3 ISO 27002

ISO 27002 is a supplementary standard that focuses on information security controls and provides best practice guidance on applying the controls listed in Annex A of ISO 27001. The ISO 27002 framework is much more cyber-focused than the ISO 27001 standard. The standard highlights how each control operates, the purpose of the control, and how to oversee the implementation. There is no certification or accreditation for ISO 27002. ISO 27002 framework documents have the following policies and points:

- *Risk Assessment*: Understand assets, their threats, and how likely the threat can successfully be used to exploit an asset
- *Security Policy*: Formal document outlining what is required when implementing the system(s)
- *Organization of Information Security*: Details how authorized staff focus on data security

- *Asset Management*: An inventory and classification of assets details
- *Human Resource Security*: Details of the management around the lifecycle of employees, e.g., the security of personnel joining and leaving an organization
- *Physical security*: Managing and limiting access to physical systems including perimeters and facilities
- *Communication and Operations*: Technical operations-based security, e.g., network systems and firewalls, Internet front doors
- *Access Control*: Management and securing of access to infrastructure
- *Information Systems Acquisition, Development, and Maintenance*: Security from the ground up
- *Incident Management*: Security incidents and related processes and procedures around cybersecurity
- *Business Continuity Management*: Business-critical functions and protecting these
- *Compliance*: Complying with standards, rules, and regulations and applicable laws

11.4.4 ISO 38500

This standard guides advising, informing, or assisting directors where a director may be any of the organization's senior members, external, technical, legal, and professional bodies. The standard also guides those advising, informing, or assisting governing bodies including executive managers, members of groups monitoring the resources within the organization, external business or technical specialists, internal and external service providers, and auditors [12]. A principal advantage of the ISO 38500 IT governance framework is to ensure that accountability is assigned for all IT risks and activities. The objective of this standard is to provide a framework of principles for directors to use when *evaluating*, *directing*, and *monitoring* the information technology in the organization. The standard is applicable for both large and small industries in the ICT space in IoE networks. The standard is applicable across all organizations including public and private companies and government entities which use IT. The standard strives to promote effective and efficient IT services in organizations through the following:

- Building stakeholders' confidence on organizing IT governance
- Guiding governing bodies about use of IT in the organization
- Establishing familiarity with the principles of the governance of IT
- This standard context consists of five elements:
  - Source of authority
  - Regulatory obligations
  - Business pressure
  - Stakeholder expectations
  - Business needs

11.4.5 HIPAA

The Health Insurance Portability and Accountability Act (HIPAA) is a set of standards (and laws) designed to inform organizations how they need to deal with and protect their systems concerning the healthcare information contained in them [13]. It is predominantly focused on the privacy of the data, but compliance with HIPAA is designed to ensure the CIA of PHI is maintained.

11.4.6 GDPR

GDPR is intended to cover the data privacy both in Europe and outside the EU through ensuring that any company who collects the data anywhere in the world must agree to comply with GDPR before being allowed get the data.

11.4.7 PCI-DSS

This is an industry standard as opposed to something defined and controlled by the government. Payment Card Industry Data Security Standard (PCI-DSS) is the result of an alliance of several credit card companies to ensure the safe, standardized handling of credit card data. It is not a law or regulation; it is self-imposed by the industry. Most small stakeholders get around their PCI requirements by using a PCI-compliant third-party provider. The PCI DSS is a collection of security standards governed by the Payment Card Industry Security Standards Council (PCI-SSC) [14]. This framework has been designed to secure credit and debit card transactions against data theft. PCI-DSS is a requirement for any organization that processes credit or debit card transactions. PCI certification is also considered the best way to safeguard sensitive data and information for card processing organizations.

PCI requires that all level 1 businesses (those organizations processing more than six million credit card transactions per year) undergo a yearly PCI audit conducted by a qualified auditor. PCI issued version 4.0 on March 31, 2022. The PCI DSS is a global standard that establishes a baseline of technical and operational standards for protecting financial account data. PCI-DSS v4.0 replaces the current PCI-DSS version 3.2 standard. Failure to comply with PCI-DSS means organizations will face huge financial penalties, damage to the company's reputation, and a loss of customer trust. Complying with PCI-DSS is a must for card processing organizations.

11.4.8 NIST-800-53

This standard mainly concentrates on privacy and controls in information systems and organizations aiming to secure assets, individuals, and operations in organizations from different cyber-threats, including human error, hostile attacks, failures in structure, natural disasters, privacy risks, and threats from foreign intelligence entities.

11.4.9 COBIT

The Control Objectives for Information and Related Technologies (COBIT) framework was developed in 1993 by ISACA and has been revised several times with COBIT-5 (2012) now the current standard. COBIT is an international governance framework and is extensive. COBIT-5 certification is available [15]. COBIT-5 is globally accepted through its use of a common language with a focus on communication among all stakeholders. The COBIT framework aims to help organizations to create a governance system that is flexible and tailorable. COBIT describes how IT tasks can be positioned into generic processes and control objectives. Cybersecurity is only one of the many parts of this IT governance.

Although COBIT is large and complex, it does provide a common language for IT professionals, stakeholders, and management. COBIT 5 does an emphasis on information security. This aids organizations meet their business challenges, especially in areas of regulatory compliance, risk management, and lining up IT strategy with organizational goals. COBIT-5 is based on five principles that are essential for the effective management and governance of enterprise IT as follows:

- *Meeting stakeholder needs* – All operations and processes should be directed toward achieving business objectives and more.
- *Covering the enterprise end-to-end* – Creating value through governance and assigning roles and responsibilities …
- Applying a single integrated framework throughout
- *Enabling a holistic approach* – Allowing for greater organizational collaboration and achievement of common goals
- *Separating governance from management* – COBIT-5 firmly believes that activities and responsibilities must be differentiated, because each serves a different purpose.

These five COBIT principles sit on a foundation of seven COBIT enablers. These are to enable the organization to build a holistic framework for the governance and management of IT. In addition, COBIT also defines 37 processes which are further grouped into 5 domains:

- *APO* – Align, Plan, and Organize
- *EDM* – Evaluate, Direct, and Monitor

- *BAI* – Build, Acquire, and implement
- *DSS* – Deliver, Service, and Support
- *MEA* – Monitor, Evaluate, and Access

11.4.10 PRINCE2

The Projects In Controlled Environment (PRINCE2) standard is a *generic project management standard* widely used for managing software projects. PRINCE2 is widely understood and recognized, and there is PRINCE2 accreditation. PRINCE2 specifies what needs to be done rather than how to do it. It claims to be the recipe for the perfect project and also that it can be tailored for any project which can result in it having a very broad scope, defining all and nothing. PRINCE2 has a strong focus on feedback and attempts to be very flexible providing a common vocabulary. PRINCE2 also claims to promote consistency of project work and the ability to reuse project assets. PRINCE2 does not provide any specialist aspect although it's broad; it's not focused on any specific industry and does not provide any leadership capability, nor does it provide specific cybersecurity guidance. PRINCE2 defines a structure of principles, themes, processes, and environment.

11.4.11 NIST CSF

The Cyber Security Framework (CSF) framework was developed in 2013 and 2014 by NIST. It is the US-based National Institute of Standards and Technology, a non-regulatory section of the US Government. The aim was to help businesses to manage and mitigate cybersecurity risks. NIST has many similarities to ISO 270001 but there are no audits. NIST's CSF is a developing document. NIST frameworks are designed to be flexible and voluntary with a strong industry focus. The focus is intended to help the industry mitigate cybersecurity risks for critical infrastructure. NIST is primarily aimed at IT in the USA and aims to have a low adoption cost, but the CSF is used by organizations and governments around the world. NIST has five core components with further subdivisions into sub-categories. These components are identifying, protecting, detecting, responding, and recovering.

NIST CSF is based on some beliefs that workers outside the security team do not understand cyber-risk and therefore fail to "own" critical mitigation tasks and also how to address risk items and (lack of) knowledge of current tools and what's available in the marketplace. NIST offers the CSF as a set of optional standards, best practices, and recommendations for improving cybersecurity and risk management in the organization.

11.5 Technical Comparison of the Standards

Table 11.1 shows a comparison of cybersecurity standards with consideration of the concepts/attributes which are auditable, cost/effort to implement, targets in terms of cybersecurity, and broadness across the organization. These are the standard affecting the whole organization or quite specifically. The main focus of these standards is covering IoE organizations like industry and whether broad or narrow, IT or CSF, and global or local level networks and organizations. Table 11.1 shows the technical comparison of discussed standards.

11.6 A Security Framework for IoE Networks

As with standards, no one framework covers all aspects of an IoE network and risk requirements. However, to choose a framework, we must first understand what one is and why we might choose it. A cybersecurity framework is a set of best practices, standards, and recommendations that help an organization protect itself from cybersecurity risks. These frameworks guide organizations to implement and meet standard requirements, and by meeting those requirements and implementing the standards, they protect their data. The NIST cybersecurity framework was designed to fill a gap in standards when it comes to cybersecurity. Differing sets of standards, policies, and guidelines in the area have meant that cyber-criminals have been successful in exploiting the many vulnerabilities the gaps the policies and standards have left. NIST aims to collectively tackle the problem with a set of well-defined uniform standards and guidelines aimed to close the gaps and standardize the controls to mitigate the risk. NIST-CSF gives a comprehensive set of guidelines and tools to help you implement a cybersecurity program for the IoE network. The framework is organized into an easily understood set of five key functions.

1. *Identify*: Develop an organizational understanding to manage cybersecurity risk for systems, assets, data, and capabilities.
2. *Protect*: Develop and implement the appropriate safeguards to ensure delivery of services.
3. *Detect*: Develop and implement the appropriate activities to identify the occurrence of a cybersecurity event.
4. *Respond*: Develop and implement the appropriate activities to action regarding a detected cybersecurity event.
5. *Recover*: Develop and implement the appropriate activities to maintain plans for resilience and to restore any capabilities or services that were impaired due to a cyber-attack.

As well as giving the guidelines, the framework provides tools to help the organization do the following:

Table 11.1 Technical comparison of standards

–	Primary objective	Type of policy	Standard used/ implemented
ISO 27001	Designed to build the foundations of information security Goal: To plan, implement, operate, monitor, and improve the ISMS There are 114 Annex A controls, divided into 14 categories A significant difference between ISO 27001 and COBIT is ISO 27001 is aimed specifically for information security, whereas COBIT is aimed for management and governance of information technology related business processes	ISMS scope document Information security policy Risk assessment process definition Statement of applicability (whether a control from Annex is applicable) Risk treatment process Information security policy Mobile device policy Remote access/teleworking policy Access control policy Cryptography policy Cryptography key management policy Clear desk and screen policy Acceptable use of information Assets policy Communications (information transfer) policy Secure development policy or plan Supplier management security policy	Global standard Framework which is auditable and touches most aspects of an organization, that is, the standard is broad in its reach ISO certification is valid for 3 years after which a recertification audit needs to be carried out Companies are required to perform surveillance audits for 2 years, and in year 3, a recertification audit is required
ISO 27002	Designed to implement controls and security management Much more cyber focused than ISO 27001 The ISO 27002 framework provides best-practice guidance on applying the controls listed in Annex A of ISO 27001	Cybersecurity Security policy – formal document outlining what is expected when implementing systems The policy offers guidance on the selection, implementation, and management of security controls based on the organization's information security risk environment	Global standard Implemented in the organization's IT/ cybersecurity space The standard is not auditable Process: Identify risks to an organization's information Implement controls appropriate to risks Monitor the organization's performance

(continued)

Table 11.1 (continued)

–	Primary objective	Type of policy	Standard used/ implemented
ISO 38500	A standard which indicates how an organization should evaluate, direct, and monitor their information technology Not as comprehensive as COBIT but aimed at senior management and also auditors	To promote effective and efficient IT services in organizations through: Building stakeholders confidence on organizing IT governance Guiding governing bodies about use of IT in the organization Establishing familiarity with the principles of the governance of IT	Global standard, based on the Australian Standard AS 8015 Many similarities to ISO 27002, implemented in the organization's IT/ cybersecurity space
COBIT-5	COBIT is an international governance framework and structures IT tasks into generic process and control objectives Focuses on management of information technology and governance Cybersecurity is only one part of the IT governance	COBIT 5 key principles: Applying a single integrated framework Meeting the stakeholder needs Covering the enterprise from end-to-end Enabling a holistic approach Separating governance from management 7 COBIT enablers: People, policies, and frameworks People, skills, and competencies Culture, ethics, and behavior Processes Organizational structures Services, infrastructure, and applications Information COBIT 5 defines 37 processes which are grouped in 5 domains: APO – Align, Plan, and Organize BAI – Build, Acquire, and Implement DSS – Deliver, Service, and Support EDM – Evaluate, Direct, and Monitor MEA – Monitor, Evaluate, and Assess	Global Standard. COBIT is an international governance framework and is very well known. COBIT-5 is very broad and touches all areas of the organization

(continued)

Table 11.1 (continued)

–	Primary objective	Type of policy	Standard used/ implemented
PRINCE2	Focus on project management and delivery with feedback Aimed at stakeholders who would likely be senior level managers. There is little focus on cybersecurity – primary focus is on project management	Structured project management with 7 principles but broad and can be used in many areas of the industry. 7 PRINCE2 principles: Continued business justification Learn from experience Defined roles and responsibilities Manage by stages Manage by exception Focus on products Tailor to suit the project environment	Developed originally in the UK as a government standard but now in wider use Used in the UK, Western European countries, and Australia As this is a project management framework, it can be used across the organization where desired There is a PRINCE2 practitioner certification programmer

- Create a risk profile to determine the organization's current level of cybersecurity risk.
- Identify the relevant standards to improve the controls and measures the organization puts in place.
- Help the organization develop new cybersecurity initiatives and requirements.
- Communicate the initiatives throughout the organization.

11.7 Conclusion

IoE networks paradigm emerged with new businesses, industries, and people's everyday routine processes. These networks are heterogeneous in nature and connected with backbone, cloud, and edge computing infrastructure. Due to these networks' complex nature, security threats and attacks are more serious concerns for these networks. The existing cybersecurity frameworks and standards are used in these networks to protect the user data and network. However, the existing cybersecurity standards still need improvements in many aspects. This chapter discussed the existing standard such as ISO 27002, ISO 38500, COBIT/COBIT 5, PRINCE2, and NIST CSF. Although these standards are adopted for backbone networks and by organizations and industries to fulfill security requirements, still there is need to develop more specific standards or frameworks to deal with these networks. This chapter also discusses these standards and compares all technically to examine their features and weaknesses. In last, the chapter also suggested the standard framework and main points to design a more feasible standard for IoE networks.

References

1. Qureshi KN, Jeon G, Piccialli F (2020) Anomaly detection and trust authority in artificial intelligence and cloud computing. Computer Networks:107647
2. Jagatheesaperumal SK, Rahouti M (2022) Building Digital Twins of Cyber Physical Systems With Metaverse for Industry 5.0 and Beyond. IT Professional 24 (6):34–40. doi:https://doi.org/10.1109/MITP.2022.3225064
3. Radanliev P, De Roure D, Nurse JR, Nicolescu R, Huth M, Cannady S, Montalvo RM (2019) New developments in Cyber Physical Systems, the Internet of Things and the Digital Economy–discussion on future developments in the Industrial Internet of Things and Industry 4.0. doi:https://doi.org/10.1007/s00146-020-01049-0
4. Zhan J, Dong S, Hu W (2022) IoE-supported smart logistics network communication with optimization and security. Sustainable Energy Technologies and Assessments 52:102052. doi:https://doi.org/10.1016/j.seta.2022.102052
5. Rehman M, Javed IT, Qureshi KN, Margaria T, Jeon G (2022) A Cyber Secure Medical Management System by Using Blockchain. IEEE Transactions on Computational Social Systems:1-14. doi:https://doi.org/10.1109/TCSS.2022.3215455
6. Williams BR, Adamson J (2022) PCI Compliance: Understand and implement effective PCI data security standard compliance. CRC Press,
7. Nah E-H, Cho S, Kim S, Cho H-I, Stingu C-S, Eschrich K, Thiel J, Borgmann T, Schaumann R, Rodloff AC (2017) International Organization for Standardization (ISO) 15189. Annals of laboratory medicine 37 (5):365–370. doi:https://doi.org/10.1128/9781555817282.ch22
8. Mahmud SY, Acharya A, Andow B, Enck W, Reaves B Cardpliance: PCI DSS compliance of android applications. In: Proceedings of the 29th USENIX Conference on Security Symposium, 2020. pp 1517–1533
9. Abdalla RS, Mahbub SA, Mokhtar RA, Ali ES, Saeed RA (2021) 6 IoE Design Principles and. Internet of Energy for Smart Cities: Machine Learning Models and Techniques:145
10. Leite JRE, Ursini EL, Chmielewski AMM, da Silva AJD New Technological Waves Emerging in Digital Transformation: Internet of Things IoT/IoE, 5G/6G Mobile Networks and Industries 4.0/5.0. In: Proceedings of the 8th Brazilian Technology Symposium (BTSym'22) Emerging Trends and Challenges in Technology, 2023. Springer, pp 329–339
11. Alshar'e M (2023) CYBER SECURITY FRAMEWORK SELECTION: COMPARISION OF NIST AND ISO27001. Applied computing Journal:245–255. doi:https://doi.org/10.52098/acj.202364
12. Rama AK, Gunawan E Evaluation of IT Governance Implementation Using COBIT 5 Framework and ISO 38500 at Telecommunication Industries. In: 2020 International Conference on Information Management and Technology (ICIMTech), 2020. IEEE, pp 453–457. doi:https://doi.org/10.1109/ICIMTech50083.2020.9211275
13. Shachar C (2022) HIPAA, privacy, and reproductive rights in a post-Roe era. JAMA 328 (5):417–418. doi:https://doi.org/10.1001/jama.2022.12510
14. Seaman J (2020) PCI DSS: An integrated data security standard guide. Apress,
15. Fernandes AJ, Hartono H, Aziza C (2020) Assessment IT governance of human resources information system using COBIT 5. International Journal of Open Information Technologies 8 (4):59–63

Chapter 12
Future Privacy and Trust Challenges for IoE Networks

Abeer Iftikhar and Kashif Naseer Qureshi

12.1 Overview

Internet of Everything (IoE) is a recent trend that is more prominent specifically for smart homes. IoE device adoption in homes is rising due to the quick and smart processes and number of services. IoE has enhanced the Internet of Things (IoT)-capable devices equipped with specialized sensors especially improvised philosophical transactions and control of services. These services are boosting operational efficiency, offering novel professional opportunities, and improving the quality of life. The absence of protective regulations and preventive controls is a clear risk to these devices for quick adoption and maintaining end nodes' or users' privacy and security against disruptive attacks designed to incur financial losses. This chapter explores the recent privacy and security challenges for IoE networks and proposed a conceptual model based on blockchain and artificial intelligence (AI) methods.

12.2 Internet of Everything

Industries are increasingly using Internet of Everything (IoE) technologies to update their operations. These companies are vulnerable to risks and security breaches due to very distinctive characteristics of such settings, particularly their sensitive

A. Iftikhar
Department of Computer Science, Bahria University, Islamabad, Pakistan

K. N. Qureshi (✉)
Department of Electronic & Computer Engineering, University of Limerick (UL), Limerick, Ireland
e-mail: kashifnaseer.qureshi@ul.ie

K. Naseer Qureshi et al. (eds.), *Cybersecurity Vigilance and Security Engineering of Internet of Everything*, Internet of Things,
https://doi.org/10.1007/978-3-031-45162-1_12

transmitted data and the frailness of the linked objects. The major objective of our endeavors is to develop a cybersecurity plan that can manage any risks that have an impact on an IoE environment while remaining within the allocated budget. By permitting the selection of a portfolio of security controls that lowers direct expenses while maximizing security level control, a financial strategy based on portfolio management is employed to achieve this goal [1]. Robust optimization is used to solve the problem's uncertainty, and it evaluates all potential risks that an attacker may create across the IoE environment using the min-max criteria. To address the problem, we employ a novel iterative approach while working under restrictions, and we evaluate its effectiveness against the Non-Dominated Sorting Genetic Algorithm (NSGA-II) meta-heuristic. By identifying effective Pareto fronts for the two investigated objective functions, the quantitative findings found. while evaluating the performance of the suggested approach, its efficacy. Our solution, which is based on the iterative method, beats the genetic algorithm by providing acceptable results for a range of issue sizes while upholding cardinality restrictions in a fair amount of time [2].

IoE includes procedures, people, data, and IoT. IoE is based on the idea of IoT, which is concerned with tying together network devices with specialized sensors across the Internet. The sensors are capable of recognizing and reacting to changes in their immediate environment, including light, temperature, sound, vibration, and others. IoE greatly expands the capabilities of IoT by including elements that might produce even better experiences for businesses, people, and governments. For instance, IoE could use all relevant data and procedures to make IoT more relevant and advantageous to humans, as opposed to relying solely on objects to interact with their environment [3].

Despite IoE connecting trillion users, devices, systems, objects, and interfaces for autonomous Internet-based services, it suffers from implementation issues and vulnerabilities of security and privacy along with architectural/infrastructural considerations. IoE is designed to benefit valuable users creating compound impacts through close and handy interconnectivity and interoperability among processes, things, data, systems, institutes, and individuals over heterogeneous platforms establishing seamless transactions incurring all fields. IoE has posed substantial security threats to its users and adopters due to the significant growth of IoE-based devices encompassing processes, users, and IoT-based mechanisms. In this chapter, we will highlight, comprehend, and investigate various IoE enabling technologies, architectural mechanisms, potentials, and outlooks for its effectual realization keeping the vital security-, trust-, and privacy-related challenges, issues, and countermeasures [4]. This research will identify and suggest a roadmap and a way forward against various attack scenarios and countermeasures against prominent cyber and network attacks like DDoS, DoS, Badmouthing, Sybil, etc. which hamper intelligent devices and gadgets operating in IoE systems. Further, current challenges, their countermeasures, and future research directions are discussed. The key insights essential for the futuristic implementation of IoE systems with tangible and concrete solutions to alert and prevent the IoE users of imminent security attacks and threat perspective through actionable steps for threat identification, recognition, and

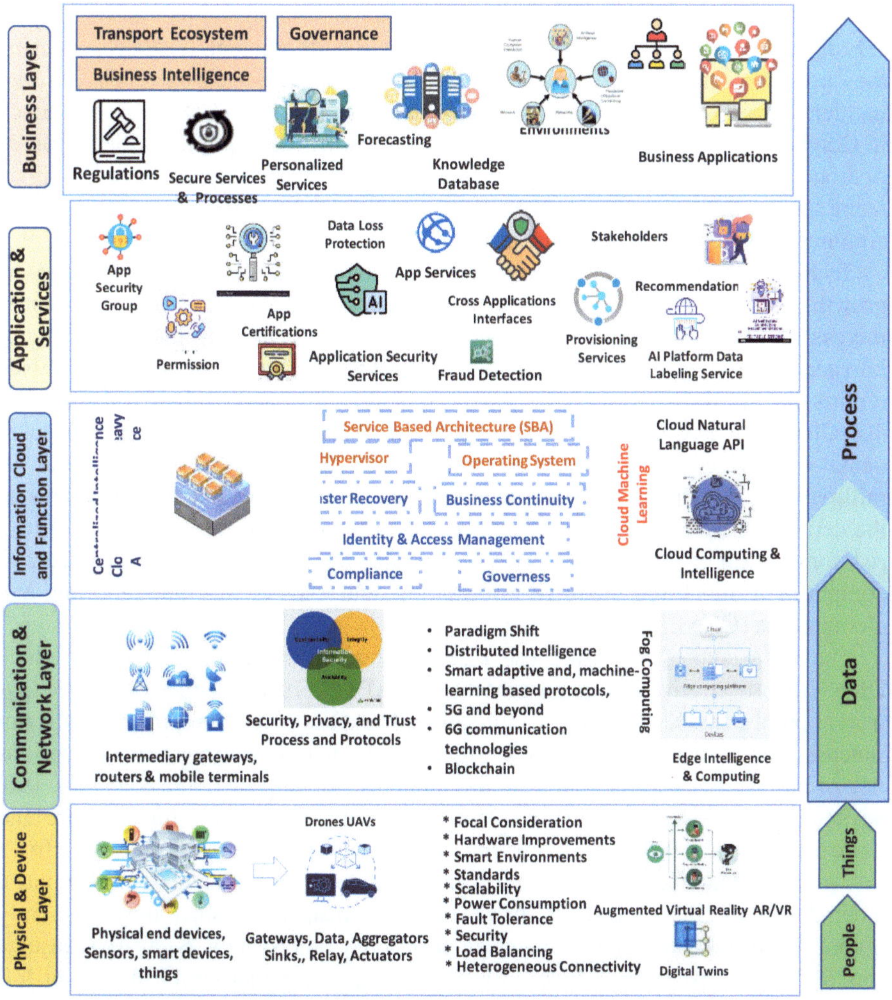

Fig. 12.1 IoE layered infrastructure

mitigation through device-friendly threat models jointly dealing with people, data, and processes fitting together resulting in formidable security threats.

IoE may provide significant security threats to its users despite all of its apparent advantages. IoE is depicted in Fig. 12.1, and the number of IoE devices in our environment is rising quickly. The IoE is playing a more significant role in our daily lives. IoE establishes a connection between cyberspace and the physical world, in particular, increasing the risk of cyberattacks aimed at IoE devices because such attacks may directly affect the health and welfare of IoE's end users. It is easy to imagine a threat scenario where an attacker purposely produces a gas leak using our gas valve example as a reference.

The fact that we frequently are unaware of the enormous number and variety of IoE devices that are present all around us, let alone the potential security dangers they provide, is even more concerning. This view is supported by recent security incidents brought on by IoE security weaknesses. One of these was a DDoS attack in October 2016 against Dyn. This event included the Mirai botnet, which consists of around 100,000 IoE hosts, including routers and digital cameras. The Dyn domain name servers failed as a result of DDoS attacks from the Mirai botnet (DNS). Major commercial websites went offline as a result (e.g., Netflix and CNN).

Technology advancements and modernizations are the key drivers of economic growth, but they also increase cyberattacks often, and violations are continually increasing. While investing in cybersecurity, it is imperative to take cyberattacks into account as they encompass all risks to wireless IoE information systems. Cisco reports that during 2018 and 2019, attacks including the bandwidth of 100 Gbps and 400 Gbps increased by 76%, for instance, there have been 15.4 million distributed denial-of-service (DDoS) attacks worldwide in 2023, a threefold increase from 7.9 million in 2017. Clients' personal and professional assets are always dependent on the dependability and security of their wireless Internet connectivity. The security risk for consumers and organizations is increased by the use of digital media, cloud computing, critical and sensitive information analysis, AI, machine learning, and e-commerce (Cisco, 2018).

The risks connected with these threats' ongoing evolution currently include DDoS, extortion, advanced persistent threats (APT), viruses, worms, malware, spyware, botnets, spam, identity theft, phishing, hacktivism, and the prospect of state-sanctioned cyberwarfare. Failures of critical infrastructure continue to pose a serious risk to almost all operators. Attacks are launched by amplified attackers who take advantage of wireless network weaknesses. Organizations now find it more and more difficult to be fully updated on the most recent security threats due to the IoT's fast growth. Manufacturers generally concentrate on utility and remote control when designing linked products' however, this is insufficient since they frequently ignore security flaws in the architecture or design of the system. Security measures should be established to reduce all potential exposures that could impair an IoE ecosystem to reduce the costs associated with a cyberattack [5]. The decision-makers implement these cybersecurity measures to defend the system from intruders. The use of shoddy authentication and encryption protocols, which makes these related things susceptible to data theft, is one of the key issues. Device systems could perhaps be "closed," which would make remote maintenance and update tough.

This is a fundamental factor to take into account when talking about IoE because it makes it timid from an operational sense to physically interact with each intelligent device and address issues. Furthermore, suppliers should implement access control procedures, protect specific hardware components, store sensitive data at different locations, and enhance employee security training to protect fragile utilities concerning wireless support and key components of the wireless IT infrastructure and physical structures. Making better use of the budget is a critical concern for decision-makers, and achieving this goal requires developing a cybersecurity investment plan. Before investing in cybersecurity, the necessary investment budget must be carefully investigated and justified.

With such a wide range of security controls that all provide security against the exploitation of overlapping vulnerabilities to some extent, selecting the best cybersecurity strategy can be challenging. To avoid and foresee dangers in the IoE environment, a solid cybersecurity investment plan must be developed, while costs are kept low. We take into account a direct cost, which consists of operational expenditures (Opex) for maintaining security controls and capital expenditures (Capex) for their purchase and installation. Remembering that the knapsack issue is NP-hard in the realm of combinatorial optimization goal of this final one is to decide what items should go in the bag for a given weight to maximize the number of usable objects while staying within the weight limit.

Economic growth is primarily driven by technological developments, but they also make cyberattacks more frequent, and transgressions are still rising. While investing in cybersecurity, it is crucial to consider cyberattacks as they encompass all risks to wireless IoE information systems. Cyberattacks are divided into three groups: the inner grouping of basic assaults, the middle grouping of malware attacks, and the outer grouping of more advanced or complicated attacks [6]. DDoS attacks will triple in number from 7.9 million in 2018 to 15.4 million in 2023, according to Cisco, with a 76% increase in attacks between 100 Gbps and 400 Gbps from 2018 to 2019 [7]. Customers depend on the dependability and accessibility of their wireless Internet access to safeguard their private and professional assets constantly. Clients as well as companies face greater security risks as a result of e-commerce, mobile payments, cloud computing, vital data analysis, artificial intelligence, machine learning, and interactive technologies [8]. In addition to state-sponsored cyberwarfare, the list also covers DDoS attacks, ransomware, Trojans, viruses, worms, malware, spyware, spammers, and spam. Failures of vital infrastructure continue to be a substantial hazard for nearly all operators. Attackers who are more powerful target the flaws in wireless networks [9].

Due to the IoT's rapid growth, businesses have a harder time staying current on the most recent security risks. When designing a connected product, manufacturers typically concentrate on functionality and remote control, but this is insufficient because they frequently ignore security flaws in the system's design or architecture [10]. Security measures should be implemented to eliminate any kind of vulnerabilities that might hurt an IoE ecosystem to reduce the costs related to a cyberattack. To protect the system from attacks, the decision-maker adopts certain cybersecurity measures. The adoption of weak encryption and authentication methods, which leaves these connected gadgets vulnerable to data theft, is one of the main problems [11]. It might be challenging to determine the best cybersecurity strategy since so many security solutions are available, each of which provides some level of defense against exploiting overlapping vulnerabilities. Developing a solid cybersecurity investment strategy is crucial while keeping prices reasonable to prevent and foresee threats in the IoE environment; we consider the direct cost, which is made up of the capital expenditure (Capex) for buying and installing a security control and the operating expenditure (Opex) for keeping it up to date. The main goal of this final challenge is to determine which items must be included in the package to achieve a certain weight to stay within the limit and maximize the number of IoT-based objects. This last challenge is similar to others in that it faces challenges like the problem of knapsack which is NP-hard in combinational optimization [12].

12.3 Concepts, Basic Cardinals, Significance

Cybersecurity As a term cybersecurity is defined as "a grouping of measures that can be taken to safeguard the user's assets, particularly linked computing devices, including laws, regulations, risk management strategies, training, industry standards, assurance, and technologies" like IoE. Last but not the least, cybersecurity attempts to protect organizational and user assets' security features from pertinent security risks in the cyber-environment. Security goals include (i) secrecy/confidentiality, (ii) integrity, and (iii) availability, as defined by the CIA triad. Confidentiality entails not allowing information to be unlawfully divulged to unauthorized people, processes, or devices. Protecting information from unauthorized change or destruction is called integrity [13].

Cybersecurity Threats Scholarly articles have identified several categorization techniques; its root objective is to identify and comprehend the characteristics and various sources of hazards to protect system assets [14]. However, the threat is defined as the potential for an attacker to damage a system or the actual actions an attacker takes with a system. It is a strategy used by attackers to take advantage of flaws in system components or the effects of threats on an asset. The literature has distinguished two categories of risks. The threat space may be divided into three dimensions with the label's agent, motive, and location for classification depending on attacking techniques. These models are called three orthogonal-dimensional models.

The threat cube categorization model should also be highlighted, which considers three primary variables: source, activity, and frequency. Three elements make up the pyramid model: attackers, vital areas, and losses. We use Microsoft's STRIDE model, which is built on identity spoofing and manipulation of data including tampering, repudiation, data rejection, information disclosure, DoS, and privilege elevation, to describe the threat impact domain. The following are the five main effects and services of security threats as listed in the ISO standard (ISO 7498-2): Data destruction includes a variety of actions, such as erasure, data corruption or change, theft, removal or loss of data, disclosure of data, and suspension of services. A linear threat categorization is a different approach to classification that divides threats based on their agents and instruments [15].

IoE In context-aware settings, the IoE automatically links people, data, processes, and things. These four pillars become integrated by serving specialized functions, increasing the relevance and value of networked relationships.

People The first pillar is "People," which refers to the individuals using smart devices (such as PCs, smartphones, laptops, etc.) to submit personal or professional data to an IoE system in a variety of ways, such as their preferences, job, and health. The "Internet of Everyone" will be a part of the IoE (IoP). When combined with human activities and associated data, it may be more beneficial to use IoE context-aware intelligence from human interactions and patterns.

Procedure The second pillar, "Processes," ensures that the correct data is supplied to the appropriate person or machine at the right time via the correct process. Unlike traditional application-specific processes, IoE processes are the most crucial resource for offering a universal user experience.

Data The data from linked devices and processes are included in the third pillar, "Data." With current advancements in Big Data, machine learning, and data analytics, contextualized data processing and analysis are projected to be at the forefront of IoE. Every linked user and process produce data that must be effectively gathered, categorized, categorized, and assessed. Powerful network agents will collect raw data and combine it with other information to create more usable data and then provide it to machines, computers, and people for further analysis and decision-making. IoE provides more efficient environmental control and quicker, more informed decision-making through the conversion of data into information [16].

Things The fourth pillar "Things" refers to physical entities that are connected to the Internet and one another, such as sensors, actuators, machines, consumer products, and assets. IoE items include cyber-physical and IoT systems. The advancement of intelligent things through technology has made edge computing, processing, and decision-making based on intelligence conceivable. Moving from intelligent handheld devices to intelligent cars, intelligent homes, and eventually intelligent cities will maximize the role of things in the development of the IoE ecosystem. Figure 12.1 shows the IoE layer architecture.

12.4 Challenges and Vulnerabilities

Even though IT firms provide enormous services to customers, they abuse users' private data. However, giving these firms access to personal data is a contentious issue in many areas of the world. Users' personal information was used by nations like China and South Korea to find people who may have interacted with COVID-19-infected people. Their findings indicate that personal information has a substantial influence on infection containment. Security is critical in today's decentralized infrastructure. Traditional security techniques cannot be applied effectively due to restricted energy supply and computational resource restrictions. As a result, the chances of being attacked increased. IoE scenarios have critical security concerns that regular IoT scenarios do not. IoE systems are installed in numerous remote

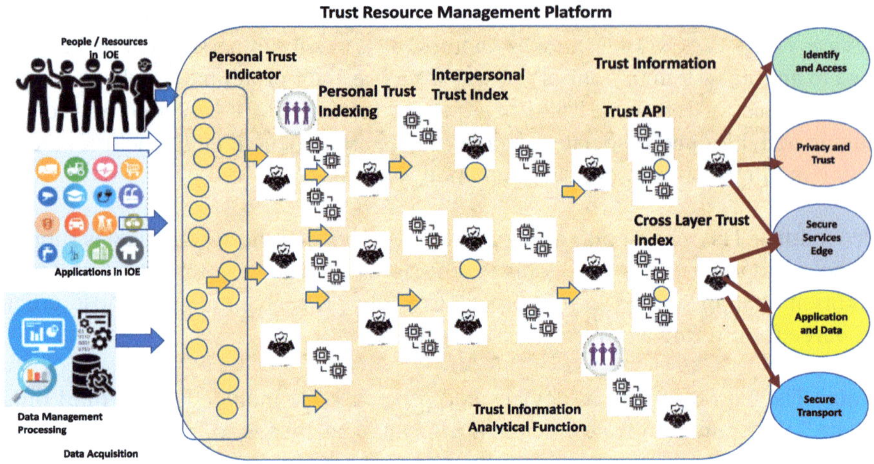

Fig. 12.2 Trust and privacy analytics in IoE-based infrastructure

locations, making them susceptible to physical attacks. Physical attacks can have catastrophic results because IoE systems depend on precise measurements from physical systems. False information affects the entire workflow of IoE control systems even though they depend mostly on data collected by physical systems [17]. Figure 12.2 shows the trust and privacy analytics in IoE-based infrastructure.

IoE has gained popularity as a result of connecting numerous devices to control physical systems in the real world, but its implementation has turned out to be a double-edged sword. There are now serious security risks as a result of this. The attack surface that attackers can employ to damage real-world systems is expanded by the Internet connectivity of microcomponents in IoE. IoE devices have unheard-of vulnerabilities since they are so accessible to hackers, have hardware flaws because of resource limitations, and have inconsistent software because there aren't enough security measures in place. We discuss the risks, openness, and hazards connected to the four essential elements of IoE in the sections that follow [18].

12.4.1 People Security

Individuals are frequently made soft targets by psychological manipulation, in addition to giving security algorithms, tools, and techniques and safeguarding key network infrastructures. Humans make mistakes and learn from them, and this process is sped up when the faults are comparable and evident. Attackers use this vulnerability to deceive individuals into making mistakes that result in catastrophic security breaches in networks. Initially, the attackers study potential victims to find possible security weaknesses. Then they win the victim's trust and give stimuli for future interactions that violate network security principles, such as phishing private information and accessing critical network resources.

Individuals are the cornerstone of network security because when people are deceived into manipulating the network, every security mechanism becomes worthless. Furthermore, human error may pose significant security concerns in networks; for example, a system is vulnerable to brute force and dictionary attacks when a system administrator uses simple and easy-to-guess password combinations. In the IoE paradigm, email, fraudulent websites, whaling, and spearfishing are tools for building people-related security risks. A virtue ethics analysis for social engineering is proposed in [19].

This study suggests measures to lower user-related security issues via thorough penetration testing. Authors in [20] proposed an approach for identifying potential workers whose information may be accessible to attackers. The attackers use this method to gain access to the whole social network of workers to discover possible network vulnerabilities. To safeguard the network from threats posed and risks faced by individuals, the scientists developed a social engineering scanner. Researchers [21] suggested that individuals improve their psychological well-being to prevent people-related attacks in today's networks.

12.4.2 Data Security

One of the most fundamental issues in the IoE paradigm is data security. Data is a valuable asset for every firm that must be protected. Robust solutions are used to protect. Furthermore, data security provides the safety of data sources, which protects data from being abused for various forms of flooding assaults. A hostile attacker continually monitoring communications in smart homes might be a data-related hazard in IoE. Although the data is encrypted, extensive measures could provide the attacker access to vital information because the creation of numerous separate identities could only be done with a small number of physical IoE devices [22].

Voting-based fault-tolerant systems face serious risks because attackers could consistently propose Sybil identities. Data security is a big concern in the massively networked IoE paradigm, where everything gathers and exchanges data over the Internet. In the IoE paradigm, secure authentication is crucial for reducing data-related attacks. Before signal transmission over the medium, device authentication should be carried out in the physical layer of IoE to ensure data security and prevent illegal data transfer. Before signal transmission over the media, device authentication should be performed at the physical layer of IoE to ensure data security and prevent illegal data access in the network layer. The most common data-related attacks are DoS and DDoS, which may be fought off via authentication [23].

In IoE, data security can be achieved through authentication techniques including key exchange, credential systems, and identity authentication and capability-based access control (IACAC). Message authentication codes may lessen man-in-the-middle (MITM) attacks. NoSQL authentication is a possible option for the IoE paradigm as it increases productivity, scalability, system performance, and deployment versatility in a variety of IoE scenarios. NoSQL may provide access controls by maintaining databases of approved devices. Data leakage, information theft, manipulation, and data repudiation are all threats from various cyberattacks. Data exchanges must be encrypted to prevent tampering and guarantee privacy and confidentiality. IoE has limited resources. Hence to be effective, data security techniques must be resource-efficient [24].

Even though RSA algorithms have been widely employed for IoT data security, there is currently not enough IoE use for them. RSA would improve data privacy and equip the system to address concerns about data leakage when used in conjunction with authentication procedures. In the IoT, hash algorithms are frequently used to evaluate the data integrity as it is sent between various nodes. To protect data against side-channel assaults, interception, and general sniffing, encryption might be used. Also, depending on domain-specific variables like available resources, transaction frequency, data rate, and desired data use, encryption could be implemented in the IoE environment. However, it is vital to tailor data security solutions to the target IoE device's resource requirements and computational power. The IoT gateway's overhead is decreased, and one of the most effective methods is using shared cryptography for communication. It uses less latency and overall network resources than other cryptography methods. Nevertheless, because of greater power consumption, its performance is a little worse than that of symmetric and public key cryptography systems.

In the IoE paradigm, it is still necessary for novel resource-efficient solutions that improve transaction security. According to recent research, hybrid encryption approaches give greater security while still maintaining efficient resource use. Furthermore, the communication devices must use the same cryptographic suites to avoid setup difficulties. Using standardized cryptographic algorithms is a strong approach for avoiding configuration concerns in data security in IoE. Considering that the IoE uses a wide range of devices and will be implemented in large-scale networks, multi-factor cryptographic solutions will be potential solutions (such as smart cities, healthcare, transportation, and aerospace, to name a few) [25].

Digital signatures effectively ensure the confidentiality and security of data transported across multiple levels and end-level devices in the IoE paradigm. Compared to AES, these approaches consume less processing power and give more efficiency than RSA. However, due to the possibility of different routing protocols being used by IoE devices, digital signatures are susceptible to domain-specific restrictions. Even though IoE traffic comes from a variety of interconnected data sources, where adversaries may also send malicious packets to learn about network configurations, traffic filtering techniques efficiently defend IoE from cyberattacks. This method circumvents the restrictions of platform-specific constraints. In the IoE environment, a prior traffic filtering approach yielded considerable benefits. Figure 12.3 shows the IoE conceptual layered essential processes for trust and privacy.

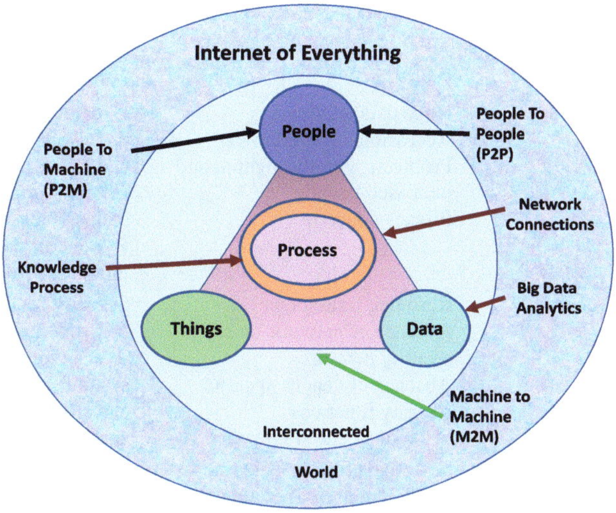

Fig. 12.3 IoE conceptual layered essential processes for trust and privacy

12.4.3 Security of Things

Safeguarding things may ensure the IoE paradigm's overall security, which faces significant security challenges due to resource limitations. Under the IoE paradigm, there may be several thing-related security considerations. But in this section, we'll examine the fundamental aspects of object security that make them vulnerable to major assaults. IoT security studies have recently been published in the literature, examining various security perspectives in the connected world [26]. Table 12.1 compares previously analyzed IoT and IoE security and trust issues that included IoT vulnerabilities, attack strategies, and empirical research demonstrating the connected world's vulnerability.

The resourcefulness of the infrastructure is a critical distinction between IoE and conventional networks. IoE is a resource-constrained infrastructure with limited memory, restricted connectivity, low power, and limited storage capacity. On the other hand, the traditional Internet contains sophisticated servers, powerful computers, smart homes, and technological devices with a wealth of resources. IoE systems require security protocols that use fewer resources and aim to balance resource consumption and security, whereas complicated and resource-intensive security solutions may maintain traditional networks with little resource consumption [27].

Table 12.1 Privacy and trust requirements and risks for IoE

S no	Challenges	Causes	Solutions
1	Storage	Huge data archiving Accessible data archiving Problems with throughput and increased latency Privacy concerns	Understanding business processes Cloud-based services Efficient privacy management Hyper converged infrastructure
2	Trust challenge	Resource limitations Absence of network-level security measures Absence of vendor-specific security functions Without a central governance	Security for certain devices Network-level remedies Access control management systems Development of security applications
3	6G in IoE	Postpone utilizing 5G systems Inefficient use of resources Absence of computationally demanding solutions Reduced bandwidth	6G's lower latency Computer-intensive solutions More bandwidth support Reduced delay
4	Value preposition	A substantial number of on-demand services Current networks lack IoT-specific QoS needs NDN's best-effort strategy Current networks don't use virtualization	Considerations for QoS unique to IoT SDN-based cache placement options Prioritizing QoS specifications Traffic routing based on QoS specifications
5	Computational complexity of IoE	Insufficient resources The higher resource needs of DL and ML The development of bottlenecks Greater use of energy	Employing edge GPUs Increasing edge GPUs' efficiency Innovative offloading techniques Cutting back on computational complexity
6	Network fault tolerant challenges	Existence of a variety of flaws Absence of fault-avoidance tactics Delivery of data in diverse networks Poor QoS data management	Making use of fault-avoidance Designed fault-avoidance techniques Fault-resilient techniques Examination of flaws
7	Scalability	Ability to connect billions of devices More data production Identification of each item Both vertical and horizontal expansion	Automated booting up Managing the IoE data pipeline Multifaceted strategies Previous business experience

12.5 Data Trust and Mistrust in IoE

The reputation of the sensor nodes may be used to determine the reliability of data in a hybrid human-device environment like IoE. Maintaining trust is a significant barrier for IoE apps that access and store data. A wide range of possible security vulnerabilities are generated by distributed, real-time network settings and the variety of linked IoT devices. The security of the data that will be transferred should be addressed at the network interoperability level, and as the IoT lacks a standard design, a layer of data security would be offered by a cohesive IoT architecture. The danger of exposure may be reduced by strategies and techniques to raise user knowledge of the impact of possible IoT hazards. Potential infrastructure for the IoE in terms of data trust and mistrust is illustrated in Fig. 12.4.

The degree of veracity in knowledge assets varies between the extremes of truth and falsity. In a broad sense, the accuracy of the sensor data determines how valuable knowledge is. Security strategies must be made autonomous and self-sufficient with the least manual human interaction. Sensor and edge computing network applications require privacy, security, timeliness, relevance, completeness, and provenance assistance. The reputation of the data source reflects the source's integrity in supplying top-notch material to address shifting external requirements and situations [28]. Trust in communication and security challenges are categorized as a direct or indirect linkage of user information with linked items inside IoT landscapes. The dependability of the devices and the level of security and trust used in installing and managing the connectivity are considered when determining the trust values. There are two types of knowledge about sensors and sensor data in IoE applications:

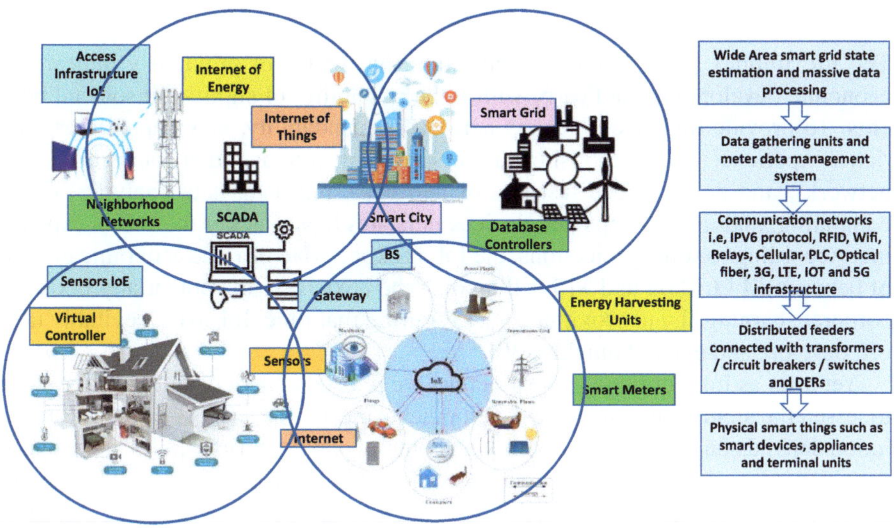

Fig. 12.4 Potential architecture for data trust and mistrust in IoE-based infrastructure

12.5.1 Trustful

Established on safeguarding the privacy of both users and service providers. To ensure the prevention and protection of user privacy in the IoE environment, meaningful identity construction, the usage of trusted communication pathways, and the preservation of contextual information are necessary. The research in [55] combined identification, authentication, and authorization into a single argument: access control to address the security of IoT products and privacy concerns. Five ideas are covered under the security dimension: non-repudiation, availability, confidentiality, and access control. Many studies have looked at issues including culpability, anonymity, and moral, ethical, legal, cultural, and regional considerations [29].

12.5.2 Untruthful

Risk arises at all knowledge transformation stages due to false or deceptive data, which ultimately results in poor judgments and serious repercussions. At the lowest level, when sensor readings or raw data are acquired, incompleteness in the data occurs. Higher levels of contextual information typically show vagueness. The heterogeneity of intelligent devices and the sensed data or authentication within many trust domains are potential security threats linked with IoT data, further complicating access control choices.

12.5.3 Trust Is Critical in IoE

The IoE implies a revolution in connection that will spur previously unheard-of economic development and carry with it the possibility of significant societal benefits. To provide higher efficiency, increased dependability, new capabilities, and richer experiences, it is establishing a "connectivity economy" that connects people, networks, and devices. The fast rise of networks, big data, data analytics, cloud computing, and mobile apps and devices indicates how quickly the IoE is expanding. According to some projections, the IoE is expected to provide economic growth of between $14 trillion and $15 trillion over the next 10 years, potentially boosting corporate earnings by approximately 21% during that time. IoE will inevitably take off, develop, and expand quickly [30].

Yet even as IoE continues to proliferate, we must also understand and take steps to lessen the significant risk these developments bring. The success of the IoE revolution depends on dependable security and ongoing "public trust" in IoE. The exponential expansion of new attack vectors as more and more devices are connected to the network, creating new vulnerabilities, and the capacity to remotely bring about physical harm or death over the Internet are some of the more intriguing concerns.

SCADA systems enable remote access to and management of vital infrastructure in factories, power plants, water treatment facilities, and pipelines for oil and gas. Now, the public has confidence in the security of our vital infrastructures, but as IoE develops and the number of devices that have access to these systems increases, we must strive to ensure that new risks are minimized, and public confidence is maintained. It brings up intriguing public policy issues regarding the state's proper function in ensuring IoE. The government has a significant leadership role. Government leadership is required to bring stakeholders together in a productive manner to facilitate the intelligent development of standards and procedures to handle security, resilience, and recovery. In the US, NIST has administered the National Strategy for Trusted Identities in Cyberspace (NSTIC) [31].

Designers can conceive of numerous scenarios in which IoE vulnerabilities may be used against us; therefore, we must move quickly to safeguard the platform if we don't want to be overrun by security flaws. Because of this, it's critical to immediately implement the appropriate security technologies, protocols, and governmental regulations to safeguard IoE's integrity and uphold the public's confidence in it [32].

12.5.4 Privacy and Trust Issues

Considerations for developing privacy and trust-aware solutions stem from the reality that the majority of users are unaware of privacy and security issues. IoE introduces a completely new way for people to interact with diverse technologies in their daily lives. Interactions between people and technology lead to the flow of sensitive data, which can raise privacy and trust issues. Figure 12.5 illustrates the generic trust, security, and privacy model for the IoE applications and infrastructure.

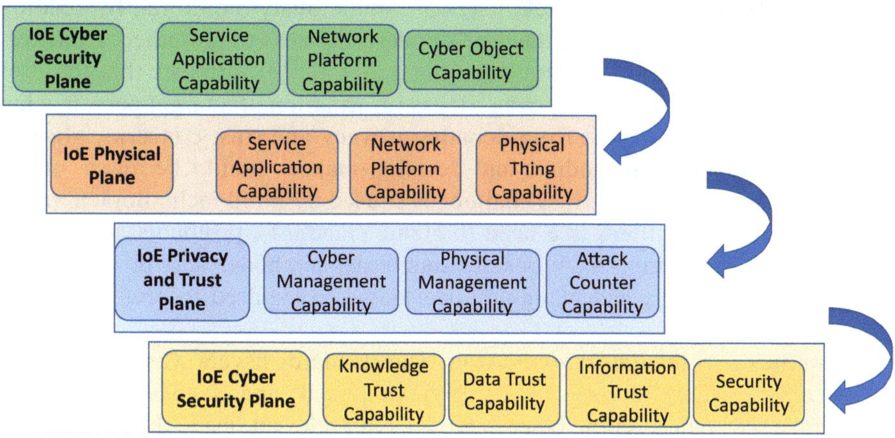

Fig. 12.5 Generic IoE trust-related capabilities and reference model

Privacy and trust difficulties in IoE can be classified into several categories, as detailed further below:

- *Identity-based trust, security, and privacy intrusion*: IoE devices collect a variety of user data that hackers might utilize for a number of illegal activities. One such activity is to link the identification with a particular person to use this data for malevolent purposes. Solutions for anonymization that isolate user identification from data should be developed to deal with these problems. Moreover, identity management and capacities for localized computing might be employed to lessen the possibility of such threats [33].
- *Location-oriented tracking*: Attackers increasingly seek to track users' whereabouts via mobile and wearable devices, and they attempt to infer sensitive information by studying user trajectories. These assaults aim to violate users' privacy and confidence. Think about the following example: A user enters a hospital and discovers that his or her location information has been falsely altered. On the basis of this sensitive information, the attacker may determine the user's health state and carry out a number of additional risky actions. Consumers must be aware of the data that sensors, wearables, and mobile devices acquire to avoid these risks. They should also have more authority and know when, how, and under what circumstances to permit and prohibit such content [34].
- *User profiling*: Attackers augment user data to create user profiles in this type of trust and privacy problem. For example, sensors in the user's surroundings may utilize personal information to determine the kind of things the user like, which could then be used to modify user attitudes toward certain products. Such information might be used to give users extra information they may not be interested in or do not want to reveal.
- *Interactions between privacy and elevation*: IoE devices might fraudulently gather private user information via microphones, cameras, keypad strokes, and user proximity. This information might be leaked and sent to the wrong people. In this type of attack, users are maliciously monitored by sending them various requests, and the user's underlying answer is studied. Attackers use such information to determine the geographical interests that the user possesses, specifically, inferring illness information of a target person using wearable sensors. Apart from debating privacy and trust concerns, governments, standardizing bodies, and decision-makers should be aware of the significant privacy and trust challenges caused by rapid technological progress. As a result, there is a need to create standards, protocols, and procedures for capturing, As a result, standards, methods, and processes for capturing, sharing, and managing user data are required to protect privacy successfully [35].
- *Standards and regulatory authorities*: The standards have a worldwide reach, ensuring privacy across national geographical and legislative borders. Regulatory bodies, on the other hand, are only relevant inside the borders of a single nation. To lessen cyberattacks, the ISO/IEC 27032:2012 Guidelines for

cybersecurity standards were developed. The main objectives of this standard are to identify best practices for Internet users' cybersecurity, lay out guidelines for reducing common cybersecurity risks, and develop a framework for stakeholder collaboration to address cybersecurity issues. It establishes four important security domains which are information security, critical infrastructure protection, network, and Internet security. ISO 27701 Privacy Information Management Systems 2019 (PIMS), ISO 29100 Privacy Framework, and ISO 27018 Protection of Personally Identifiable Information are recent ISO standards for cloud security. The ISO/IES 27701:2019 standard defines the essential requirements and best practices for developing, implementing, and evolving PIMS in an organizational environment. It expands the ISO/IEC27001 and ISO/IEC 27002 privacy management standards. The ISO 29100 Privacy Framework protects personally identifiable information (PII) inside an organization. It defines a common language, actors engaged in PII processing, privacy protection issues, and references to cutting-edge privacy concepts in IT. The ISO/IEC 27018:2019 standard establishes common control paradigms and best practices for protecting personally identifiable information (PII). This conforms to ISO/IEC 29100 for the public cloud environment. This standard has a broad scope and is relevant to organizations of all sizes and types, including public, commercial, government, and nonprofit organizations. It may be used to protect the privacy of firms that provide cloud computing services as PII processors to other enterprises.

- *The ISO/IEC 27018*: 2019 standard outlines typical control paradigms and recommended procedures for safeguarding PII. This standard is compatible with the public cloud environment specified by ISO/IEC 29100. This standard has a broad scope and is applicable to a variety of sizes and types of organizations, including nonprofit, public, and private organizations. It can be used to preserve privacy in companies that act as PII processors for other companies and offer cloud computing services. On April 14, 2016, the European Union (EU) and the European Economic Area (EEA) adopted the General Data Protection Regulation (GDPR), which took effect on May 24, 2018. This personal privacy protection law transfers power from businesses using personal data to individuals. It provides the instructions for processing personal data and is applicable to any organization handling personal data of EEA data subjects. While numerous laws to protect personal data are passed at the federal and state levels in the USA, there is no single primary data protection legislation. The Video Privacy Protection Act, the Cable Communication Policy Act, and the Driver's Privacy Protection Act of 1994 are notable pieces of legislation that safeguard privacy. The National People's Congress Standing Committee passed a law on November 7, 2016, protecting personal data, and it became effective on June 1, 2017. The privacy of personal data is protected in various countries thanks to laws like Australia's Privacy Principles, Brazil's Internet Act, Canada's Personal Information Protection and Electronic Data Act, India's Information Technology Act, South Africa's Electronic Communications and Transactions Act, and the UK's Information Commissioner's Office. Legislative bodies and privacy protection standards are directed to forbid exposing or abusing personal information. These rules' primary goal is to ensure ethical data practices in the creation, use, and archival of data. Although different standards and legislation have been suggested, putting these laws into practice is difficult due to several legislative complications, making IoE a particularly vulnerable ecosystem to privacy threats.

12.6 Security and Privacy Issues

Multiple domains exist in IoE networks by using several services and linked devices. Security, privacy, and trust standards are applied according to each domain. IoE securities now face certain specific issues, including the following:

1. *User privacy and its data protection in IoE*: Privacy is a significant concern in IoE. User privacy is a highly touchy topic in many research projects. IoE connects people, processes, data, and things; data is transmitted through the Internet. IoE network requires privacy in data gathering, sharing, administration, and security.
2. *Authentication and identity management*: IoE uses various methods and technology for authentication and identification. This aims to control and safeguard access to data and resources. Objects are identified by identity, and communication between two parties requires authentication.
3. *Trust management and policy integration*: Several items communicate in the IoE scenario, and trust is crucial to creating safe communication between them. To acquire user confidence, there should be an effective process in the IoE ecosystem.
4. *Authorization and access control*: After being recognized, authorization makes it possible to determine if the person or thing is allowed to possess the resource. Access to resources is provided or restricted based on a wide range of factors. Access controls are used to implement authorization.
5. *End-to-end security*: Security is similarly important at the locations where IoE devices link to Internet hosts. For complete end-to-end security, session keys and algorithms must be implemented securely.
6. *Attack-resistant security solution*: Various devices are connected to the IoE. Since these devices may suffer from different attacks, such as DoS, flood attacks, etc.
7. *M2M communication in IoE systems*: M2M is the technology that makes it possible for wired and wireless equipment to interact with one another. M2M connections are often used in industrial automation for machine monitoring and measurement. IoE combines people, processes, data, and things to increase the usefulness and relevance of networked relationships. First responders may benefit from IoE's pervasiveness, which can be used to accomplish many goals for many people, including M2M and IoT technologies.

12.7 Open Issues in Research, Future Trends, and Way Forward

This section outlines current problems, their root causes, and potential future study topics. These guidelines may be essential for the future successful implementation of IoE systems.

12.7.1 Challenges

- *Trust, Security, and Privacy Challenges*: The engagement of several diverse entities makes it challenging to design IoE security countermeasures, which makes putting security solutions into practice a time-consuming and tedious operation. To fully comprehend the underlying vulnerabilities of stakeholders, data, communication, people, and things before designing any solutions, a thorough analysis of all IoE-related issues should be carried out. For the smooth security of IoE, a robust security mechanism that is integrated into the architecture of the underlying system is necessary. The service-oriented architecture along with the open business model should be taken into consideration by any IoE security solution to assure the security of the underlying heterogeneous entities.
- Authors in [14] incorporated log data and graph analytic techniques to find anomalies and perform anomaly detection. Authors in [36] proposed an IoE security solution and offer useful information about the measures done to secure the IoE ecosystem. Authors in [37] provide a study proposal on examining the IoE paradigm's design criteria for user activity detection that may be considered when creating a security and privacy solution. Authors in [38] developed a unique consensus method for physical unclonable functions as part of a blockchain-based IoE solution. Although several IoE security and privacy strategies might be constructed utilizing previous IoT technologies, the security and privacy requirements should be included in the IoE paradigm from the inception [39]. This might prevent the need for upgrading security measures at a later stage of IoE development by making systems flexible. Provide an IoE security solution and offer useful information about the measures done to secure the IoE ecosystem that have been covered in Table 12.2. This table describes the underlying security approach's security settings, security components, associated issues, potential fixes, and disadvantages.
- *Storage and Serialization*: As IoE evolves with time, massive data storage will be needed. M2M, P2M, and P2P communication modules will increase data creation, necessitating efficient and practical storage solutions. Although cloud-based solutions (such as server less storage, hybrid storage, etc.) seem feasible, their cost and privacy concerns necessitate cost-effective substitutes that also address privacy concerns. In addition, there are other problems with cloud-based solutions, like latency and bandwidth. Because data will be produced continu-

Table 12.2 Trust, security, and privacy requirements for IoE

S/N	Security requirements	Descriptions
1	Confidentiality	By preventing unwanted access, the data is protected and only accessible to authorized users
2	Integrity	Maintaining end-to-end security in IoE communication and utilizing digital signatures to assure data integrity helps prevent unintentional interference
3	Availability	When users require them, data, equipment, and services must be accessible
4	Authentication	Each item within the IoE has to be able to recognize and verify other objects. In the Internet of Everything, several things, including people, services, gadgets, and processing units, communicate
5	Non-repudiation	A cybersecurity criterion that serves as evidence of entities' activities in IoE networks is no repudiation

ously under the IoE paradigm, even a minor network slowdown will make the entire process ineffective. Hence, other potential solutions must be explored, such as hyper-converged infrastructure and edge computing.

- *6G for IoE Realization*: The global standardization of 5G communication has been concluded, and implementation efforts have begun in several nations. The ramifications of 5G for mobile platforms show how it has limitations that go against its potential to be an enabling technology for Internet of Everything (IoE) applications. The requirements of cutting-edge IoE services might not be met by 5G technology, despite the fact that it can support IoE services. 6G technologies could solve the drawbacks of 5G wireless technology. People, products, processes, and data will all be able to be intelligently connected via 6G. It will change into a ubiquitous setting that can offer services with little to no human involvement. IoE communication will be transformed by 6G thanks to mobile triband dependable low latency connectivity and increased network capabilities. Real-time data analytics for ubiquity operations will be enabled thanks to the ultra-low latency. Large-scale data uploads and downloads might be done using edge computing to disseminate information among interconnected IoE infrastructures. Instead of bringing in more high-frequency bands, 6G will use wireless spectrum to bring in massive latent services and new IoE trends. Big data analytics, 6G connectivity, and ICT will all significantly increase IoE ubiquitous services. In addition to augmented reality, ubiquitous services, multi-sensor data fusion, and the merger of precision and actuation control activities, 6G will introduce a revolutionary interface for human-machine interaction [40].
- *Value Preposition in IoE*: IoE consists of traditional devices and links physical things connected by Wi-Fi, Bluetooth, or 5G beyond wireless communication. System resilience is necessary to allow and guarantee IoE services. Massive data streams may be unlocked thanks to the IoE connection, giving enterprises access to brandnew opportunities for income generation. Yet, effective communication management is crucial for an appropriate IoE execution. The procedure of select-

ing the appropriate connection is difficult. MNOs (multinational operations) provide reach across a variety of geographic locales. Hence, cooperation between various communication service providers is crucial to achieving coverage everywhere. Managing the connection complexity of the IoT requires simple, affordable, dependable, and adaptable services. With the growth of IoE, efficient IoE subscriber management is necessary. A proven new value proposition design. The winning combinations to create a successful IoE strategy include IoE business models and straightforward business strategies. IoE promises unending advantages everywhere. The best ones are better client experiences, more sales, more money, and exceptional quality. IoE's value proposition is mostly reliant on the gathering of data, analysis of data, decision-making on data utilization, and knowledge-based action. The core of the IoE value proposition is an ubiquitous operation built on data analysis [41].

- *Computational Complexity of IoE*: IoE devices have finite amounts of processing power, memory, and battery life. A lack of resources hampers the restricted use of DL and ML. A bottleneck results when intelligent applications that require a lot of computation are implemented. The current cloud computing computational offloading techniques have greater overhead energy usage. The state of the network also influences such systems' accessibility. For instance, poor network connectivity will negatively impact computation offloading, resulting in the unavailability of programs. Edge computing GPUs may be a modern remedy for DL, ML, and IoE. Unfortunately, even on limited mobile devices, powerful GPUs still use much power. Compute-intensive IoE strategies may be implemented with new offloading techniques and improved GPU-based IoE solutions. Moreover, methods that lessen IoE's computational complexity must be developed. So, a good subject for future study might be lowering computing complexity [42].
- *Fault Tolerance Trials in Networks*: IoE is an integrated paradigm with several operational flaws and is still in its evolutionary stage. An uninterrupted, fault-free operation must be ensured for IoE systems to be successfully implemented. It is difficult to ensure such a system, but solutions should be developed to create IoE systems that can continue to function well even if some of the components fail. For IoE systems to communicate the decision based on the data gathered from the connected devices, fault-tolerant systems must be developed. Failure of any IoE node could have disastrous effects because the systems won't be able to process information if any nodes are unavailable. If part of the physical systems' components fails, the linked data center won't be able to generate actionable insights in the presence of flexible and scalable systems. Most data center networks take fault tolerance into consideration, although it might be challenging to guarantee faultless operation. Yet, by utilizing thorough fault tolerance methods, faults might be prevented. To address defects in linked IoE settings, fault detection, isolation, and avoidance solutions might be created.
- *Scalability*: IoE will offer a worldwide platform for managing autonomous services that connect people, processes, things, and data. Comparing connecting billions of IoE objects to deploying and managing a few devices. In IoE systems,

scalability poses significant difficulties. Because the IoE framework will uniquely identify every linked object, it is crucial to consider any potential scaling issues arising from IoE implementation. When different systems' connection changes over time, IoE systems will need to adapt and deliver services following the needs of the underlying business process models. IoE systems must be scalable both vertically and horizontally for this. IoE has to consider the network, business, marketing, software, and hardware requirements that might be combined with IoE systems. Scalability in IoE systems must be improved using automated bootstrapping, regulating the IoE data flow, multidimensional scalability methodologies, and creating efficient microservices design. IoE architecture should be created in a way that takes future integration into account as the system is expanded. When creating scalable IoE systems, it is also essential to actively manage security, privacy, identity management, and access control. To build scalable solutions, however, addressing scalability necessitates thoroughly grasping the business processes beforehand.

The main IoE security provisioning requirements are shown in Table 12.2.

12.7.2 Open Research Issues

1. Massive data storage will be required as IoE evolves over some time. Data creation will increase with P2P, P2M, and M2M communication modules, necessitating effective and quickly available storage alternatives. Although cloud-based solutions (such as server less storage, hybrid storage, and so on) appear to be viable options, the expense of utilizing cloud-based services and privacy issues necessitates a cost-effective method that addresses privacy issues. Furthermore, cloud-based solutions have other concerns, such as latency and bandwidth. Because data is created continually under the IoE paradigm, a little bottleneck inside a network will render the entire process ineffective. Potential solutions like edge computing and hyper-converged infrastructure are required [43].
2. Given the participation of many stakeholders, designing countermeasures for IoE security poses enormous obstacles, making deploying security solutions an exhausting and tedious effort. Before establishing any solution, a thorough examination of all components of IoE should be conducted to comprehend the underlying vulnerabilities of stakeholders, data, communication, people, and objects. A strong security mechanism built into the underlying system is required for IoE security to be seamless. Any solution that caters to IoE security should consider the open business model and service-oriented architecture to secure the underlying heterogeneous entities' security. Because of the participation of many infrastructures, a rising volume of data collection and data transfer is unavoidable; consequently, data must be accessible to maintain integrity while ensuring a secure data exchange.

3. In the IoE paradigm, anomalies in big data processing might arise, endangering data integrity with vulnerable analytic approaches. Mechanisms for spotting abnormalities should be created for IoE security as a result. Authors in [44] employed log data and graph analytic methods to find anomalies. Authors in [45] designed an IoE security solution and give practical insights into the procedures required to secure the IoE ecosystem. In [46], authors considered looking at the detection of user activity design factors in the IoE paradigm, which might be leveraged to create a security and privacy solution. Authors in [38] provided a blockchain-based remedy for the Internet of Everything by offering a special consensus technique for physically irreplaceable tasks.
4. Although multiple ways to IoE security and privacy might be built utilizing existing IoT solutions, the security and privacy aspects should be integrated by design in the IoE paradigm [47]. System security would become adaptable as a result, possibly eliminating retrofitting during later stages of IoE development. Table 12.2 shows that the security and privacy criteria for developing countermeasures and security protocols include user attention, proactive security features, security by design, security by default, product lifecycle security, transparent security solutions, and security by design. This table illustrates the security requirements, security elements, participation, potential remedies, and disadvantages of using the underlying security approach [48].

IoE takes over from IoT. In the present era, security in an IoE network is crucial. The main hurdles in IoE are trust, security, and privacy concerns. To automate systems in a variety of industries, such as home automation, smart cities, smart agriculture, etc., various IoE security needs are helpful [49].

12.8 Conclusions

IoE is the networked connectivity of people, processes, things, and data, according to Cisco. By realizing the cumulative value generated from the connectivity of people, processes, data, and objects, the IoE opens up previously unimaginable prospects for individuals, communities, and nations. We have thoroughly covered a variety of IoE-related topics in terms of security, trust, and privacy in this study chapter. We outline the essential architectural elements, supporting technologies, and significant advantages of IoE. The modern use cases and synergies covered in this study guide the effective deployment of various IoE-based systems. We review the main IoE attack challenges and weaknesses and offer defenses against them. We examine solutions to IoE's security, privacy, and trust problems.

References

1. Flores FFS, de Lemos Meira SR Ethical Software Engineering: A critical review about Software Engineering in face of Security Requirements in the IoT/IoE Society. In: 2021 IEEE International Systems Conference (SysCon), 2021. IEEE, pp 1–8
2. Ilgi GS, Ever YK (2020) Critical analysis of security and privacy challenges for the Internet of drones: a survey. In: Drones in smart-cities. Elsevier, pp 207–214
3. ALiero MS, Qureshi KN, Pasha MF, Jeon G (2021) Smart Home Energy Management Systems in Internet of Things networks for green cities demands and services. Environmental Technology & Innovation:101443. doi:https://doi.org/10.1016/j.eti.2021.101443
4. Qureshi KN, Alhudhaif A, Hussain A, Iqbal S, Jeon G (2021) Trust aware energy management system for smart homes appliances. Computers & Electrical Engineering:107641. doi:https://doi.org/10.1016/j.compeleceng.2021.107641
5. Qureshi KN, Alhudhaif A, Haider SW, Majeed S, Jeon G (2022) Secure Data Communication for Wireless Mobile Nodes in Intelligent Transportation Systems. Microprocessors and Microsystems:104501. doi:https://doi.org/10.1016/j.micpro.2022.104501
6. Mazzoccoli A, Naldi M (2022) An Overview of Security Breach Probability Models. Risks 10 (11):220
7. Cisco U (2021) Cisco annual internet report (2018–2023) white paper. 2020. Acessado em 10 (01)
8. Gabilondo García Á, Fernández Z, Viola R, Martín Á, Zorrilla M, Angueira Buceta P, Montalbán Sánchez J (2022) Traffic Classification for Network Slicing in Mobile Networks.
9. Luiijf E, Klaver M (2021) Analysis and lessons identified on critical infrastructures and dependencies from an empirical data set. International Journal of Critical Infrastructure Protection 35:100471
10. Karie NM, Sahri NM, Yang W, Valli C, Kebande VR (2021) A review of security standards and frameworks for IoT-based smart environments. IEEE Access
11. Kong X, Wu Y, Wang H, Xia F (2022) Edge Computing for Internet of Everything: A Survey. IEEE Internet of Things Journal 9 (23):23472–23485
12. Shahzad K, Aseeri AO, Shah MA (2022) A Blockchain-Based Authentication Solution for 6G Communication Security in Tactile Networks. Electronics 11 (9):1374
13. Un Nisa K, Alhudhaif A, Qureshi KN, Hadi HJ, Jeon G (2022) Security Provision for Protecting Intelligent Sensors and Zero Touch Devices by using Blockchain Method for the Smart Cities. Microprocessors and Microsystems:104503. doi:https://doi.org/10.1016/j.micpro.2022.104503
14. Conti M, Kumar G, Nerurkar P, Saha R, Vigneri L (2022) A survey on security challenges and solutions in the IOTA. Journal of Network Computer Applications:103383
15. Paleri P (2022) Threat, Risk and Uncertainty: Triad of Chaotic Balance in a Chancy, Chancy World. In: Revisiting National Security. Springer, pp 155–193
16. Qureshi KN, Ahmad A, Piccialli F, Casolla G, Jeon G (2020) Nature-inspired algorithm-based secure data dissemination framework for smart city networks. Neural Computing and Applications. doi:https://doi.org/10.1007/s00521-020-04900-z
17. Sachdev R Towards security and privacy for edge AI in IoT/IoE based digital marketing environments. In: 2020 Fifth International Conference on Fog and Mobile Edge Computing (FMEC), 2020. IEEE, pp 341–346
18. El-Masri M, Hussain EMA (2021) Blockchain as a mean to secure Internet of Things ecosystems–a systematic literature review. Journal of Enterprise Information Management 34 (5):1371–1405
19. Hatfield JM (2019) Virtuous human hacking: The ethics of social engineering in penetration-testing. Computers Security 83:354–366
20. Formosa P, Wilson M, Richards DJC, Security (2021) A principlist framework for cybersecurity ethics. 109:102382

21. Abe N, Soltys M (2019) Deploying health campaign strategies to defend against social engineering threats. Procedia Computer Science 159:824–831
22. Shuvo MMH (2022) Edge AI: Leveraging the full potential of deep learning. In: Recent Innovations in Artificial Intelligence and Smart Applications. Springer, pp 27–46
23. Kokkonen H, Lovén L, Motlagh NH, Partala J, González-Gil A, Sola E, Angulo I, Liyanage M, Leppänen T, Nguyen T (2022) Autonomy and Intelligence in the Computing Continuum: Challenges, Enablers, and Future Directions for Orchestration. arXiv preprint arXiv:01423
24. Chinchawade AJ, Lamba OS (2021) Security and privacy challenges in internet of everything (ioe) with security requirements. Journal of Engineering Technology-Suresh Gyan Vihar University 7 (2)
25. Desai S, Alhadad R, Chilamkurti N, Mahmood A (2019) A survey of privacy preserving schemes in IoE enabled smart grid advanced metering infrastructure. Cluster Computing 22:43–69
26. Stellios I, Kotzanikolaou P, Psarakis M, Alcaraz C, Lopez J (2018) A survey of IoT-enabled cyberattacks: Assessing attack paths to critical infrastructures and services. IEEE Communications Surveys Tutorials 20 (4):3453–3495
27. Al-Garadi MA, Mohamed A, Al-Ali AK, Du X, Ali I, Guizani M (2020) A survey of machine and deep learning methods for internet of things (IoT) security. IEEE Communications Surveys Tutorials 22 (3):1646–1685
28. Shankhpal SV, Brahmananda S (2022) Systematic analysis and review of trust management schemes for IoT security. International Journal of Wireless Mobile Computing 23 (1):33–45
29. Farias da Costa VC, Oliveira L, de Souza J (2021) Internet of everything (IoE) taxonomies: A survey and a novel knowledge-based taxonomy. Sensors 21 (2):568
30. Saba T, Rehman A, Haseeb K, Alam T, Jeon G (2023) Cloud-edge load balancing distributed protocol for IoE services using swarm intelligence. Cluster Computing:1–11
31. Jimo S, Abdullah T, Jamal A IoE Security Risk Analysis in a Modern Hospital Ecosystem. In: Cybersecurity in the Age of Smart Societies: Proceedings of the 14th International Conference on Global Security, Safety and Sustainability, London, September 2022, 2023. Springer, pp 451–467
32. Juszczyk O, Shahzad K (2022) Blockchain technology for renewable energy: principles, applications and prospects. Energies 15 (13):4603
33. Alqasemi F, Al-Hagree S, Shaddad RQ, Zahary AT An IEEE Xplore Database Literature Review Concerning Internet of Everything During 2020–2021. In: 2021 International Conference on Intelligent Technology, System and Service for Internet of Everything (ITSS-IoE), 2021. IEEE, pp 1–8
34. Puthal D, Damiani E, Mohanty SP Secure and Scalable Collaborative Edge Computing using Decision Tree. In: 2022 IEEE Computer Society Annual Symposium on VLSI (ISVLSI), 2022. IEEE, pp 247–252
35. Aldawsari H, Artoli AM (2021) A Reliable Lightweight Trust Evaluation Scheme for IoT Security. International Journal of Advanced Computer Science Applications 12 (11)
36. Tanwar S, Popat A, Bhattacharya P, Gupta R, Kumar N (2022) A taxonomy of energy optimization techniques for smart cities: Architecture and future directions. Expert Systems with Applications 39 (5):e12703
37. Uribe S, Moreno F, Hernández G, Álvarez F (2022) Personalised Interaction or How We Can Improve Migrants' Experience When Using a Digital Companion Through a Mobile App. In: Information and Communications Technology in Support of Migration. Springer, pp 229–247
38. Mohanty SP, Yanambaka VP, Kougianos E, Puthal D (2020) PUFchain: A hardware-assisted blockchain for sustainable simultaneous device and data security in the internet of everything (IoE). IEEE Consumer Electronics Magazine 9 (2):8–16. doi:https://doi.org/10.1109/MCE.2019.2953758
39. Swathi GC, Kumar GK, Kumar AS (2022) Estimating Botnet Impact on IoT/IoE networks using Traffic flow Features. Computers Electrical Engineering 102:108209

40. Rustagi A, Manchanda C, Sharma N IoE: A boon & threat to the mankind. In: 2020 IEEE 9th International Conference on Communication Systems and Network Technologies (CSNT), 2020. IEEE, pp 114–119
41. Kouicem DE, Bouabdallah A, Lakhlef H An efficient architecture for trust management in IoE based systems of systems. In: 2018 13th Annual Conference on System of Systems Engineering (SoSE), 2018. IEEE, pp 138–143
42. Fortino G, Fotia L, Messina F, Rosaci D, Sarné GM (2020) Trust and reputation in the internet of things: State-of-the-art and research challenges. IEEE Access 8:60117–60125
43. Elsayed MA, Zulkernine M (2020) PredictDeep: security analytics as a service for anomaly detection and prediction. IEEE Access 8:45184–45197
44. Khan WZ, Rafique W, Haider N, Hakak S, Imran M (2022) Internet of Everything: Enabling Technologies, Applications, Security and Challenges. TechRxiv Preprint. doi:https://doi.org/10.36227/techrxiv.21341796.v1
45. Ryoo J, Kim S, Cho J, Kim H, Tjoa S, DeRobertis C IoE security threats and you. In: 2017 International Conference on Software Security and Assurance (ICSSA), 2017. IEEE, pp 13–19. doi:https://doi.org/10.1109/ICSSA.2017.28
46. Janbi N, Katib I, Albeshri A, Mehmood R (2020) Distributed artificial intelligence-as-a-service (DAIaaS) for smarter IoE and 6G environments. Sensors 20 (20):5796
47. Mohanty SP (2020) Security and Privacy by Design is Key in the Internet of Everything (IoE) Era. IEEE Consumer Electron Mag 9 (2):4–5
48. Bera B, Das AK, Obaidat MS, Vijayakumar P, Hsiao K-F, Park Y (2020) AI-enabled blockchain-based access control for malicious attacks detection and mitigation in IoE. IEEE Consumer Electronics Magazine 10 (5):82–92
49. Jameii SM, Zamirnaddafi RS, Rezabakhsh R (2022) Internet of Flying Things security: A systematic review. Concurrency Computation: Practice Experience 34 (24):e7213

Index

K. Naseer Qureshi et al. (eds.), *Cybersecurity Vigilance and Security Engineering of Internet of Everything*, Internet of Things,
https://doi.org/10.1007/978-3-031-45162-1

MIX
Papier aus verantwortungsvollen Quellen
Paper from responsible sources
FSC® C105338

If you have any concerns about our products,
you can contact us on
ProductSafety@springernature.com

In case Publisher is established outside the EU,
the EU authorized representative is:
Springer Nature Customer Service Center GmbH
Europaplatz 3, 69115 Heidelberg, Germany

Printed by Libri Plureos GmbH
in Hamburg, Germany